Vocational Skills Training for

Segmental Paver Installation

by Stephen Jones

paverschool.com

Published by:

PAVE TECH, INC.
P.O. BOX 576
Prior Lake, MN
Phone: 952-226-6400
Fax: 952-226-6406
Toll Free: 800-728-3832
pavetech.com

Published by:
PAVE TECH, INC.

ISBN 0-9717465-0-8
First Edition

Printed in USA

FOUNDATION SKILLS
Instructional Training Video
for Concrete & Clay Pavers

Advantages

- Become an SASP certified Installer
- Building block for in house training program
- Improve crew efficiencies
- Better profitability
- Structured training format
- Raises expectations for higher level of detail & quality
- Cost effective training format
- **Available in English and Spanish**

Using industry best practices, techniques, and technology, this training video will allow you to increase the quality and efficiency of your crews on all your segmental pavement projects. With the help of SASP's Founders, the industry's best paver contractors, you will also learn specialized installation techniques.

Chapter 1 — INTRODUCTION

Chapter 2 — SITE VISIT

Chapter 3 — MATERIAL LOGISTICS

Chapter 4 — LAYOUT

Chapter 5 — DEMOLITION

Chapter 6 — EXCAVATION

Chapter 7 — BASE

Chapter 8 — FINISH GRADING

Chapter 9 — EDGE RESTRAINTS

Chapter 10 — SCREEDING

Chapter 11 — LAYING PAVERS

Chapter 12 — MARKING & CUTTING

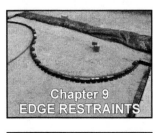

Chapter 13 — FINAL COMPACTION

Chapter 14 — LIGHTING

Certification Benefits:
- Listed as a Certified Installer on the SASP website
- Each Certification comes with a $100 voucher towards future hands-on training at SASP

Sponsors and Contributors to the Vocational Manual

Sponsors have each made a significant financial contribution to the creation of this manual. Contributors have also helped cover the costs of development and production.

Sponsors:

BIA Brick Industry Association
Leroy Danforth, Jr., EIT
Manager, Technical Publications
danforth@bia.org
11490 Commerce Park Drive
Reston, VA 20191
Phone: 703-620-0010
Fax: 703-620-3928
www.gobrick.com
brickinfo@bia.org

BORAL PAVERS

Boral Brick Company
Mark Link
1449 Doug Barnard Pkwy
Augusta, GA 30906
Phone: 706-828-7346
mark.link@boral.com

General Shale Brick
John Hammett
PO Box 3547
Johnson City, TN 37602
Phone: 423-282-4661
Fax: 423-952-4104
www.generalshale.com
jhammett@generalshale.com

Glen Gery Brick
Marketing Services Department
1166 Spring St, P.O. Box 7001
Wyomissing, PA 19619-6001
Phone: 610-374-4011
Fax: 610-374-3700
www.glengerybrick.com
gg@glengerybrick.com

Pine Hall Brick Co., Inc.
Paver Department
PO Box 1104
Winston Salem, NC 27116-1044
Phone: 800-334-8689
Fax: 336-725-3940
www.americaspremierpaver.com
paverinfo@pinehallbrick.com

Bobcat.

Bobcat Company
Attn: Infocenter
PO Box 6000
West Fargo, ND 58078-6000
Phone: 701-241-8700
Fax: 701-241-8704
www.bobcat.com
infocenter@bobcat.com

Contributors:

Acme Brick Company
2821 W. 7th
Fort Worth, TX 76107
Phone: 800-792-1234 ext 365
Fax: 817-390-2404
www.brick.com
bill@brick.com

Whitacre-Greer
Mike Longo, Exec. VP
1400 S. Mahoning Ave.
Alliance, OH 44601
Phone: 330-823-1610
Fax: 330-823-5502
Toll Free: 800-947-2837
www.wgpaver.com
mlongo@wgpaver.com

Endicott Clay Products Co.
Gary Davis
PO Box 17
Fairbury, NE 68352
Phone: 402-729-3315
Fax: 402-729-5804
www.endicott.com
endicott@endicott.com

Hanson Brick
Brickyard Road at Frost Avenue
Columbia, SC 29203
Phone: 803-786-1260
Fax: 803-786-9703
www.hansonbrick.com
Perrin.babb@hansonamerica.com

PAVE TECH ©

"This manual is dedicated to the segmental paving contractors who have persevered through the rapid growth of our industry without a road map."

Stephen Jones

CONTRACTORS from around North America and Europe have helped with input to this manual. Two of them deserve special recognition because of their service to our industry and their help and support of this manual including many out of town meetings.

BILL SCHNEIDER
LPS Pavement Co.
67 Stonehill Rd
Oswego, IL 60543
Phone: 630-551-2100
Fax: 630-551-2105
bills@lpspave.com

BOB GOOSSENS
Precise Paving, Inc
1243 45th Street
West Palm Beach, FL 33407
Phone: 561-845-6040
FAX: 561-845-1787
rgoose@precisepaving.com

I will also mention others that have been of help and support over the years. There have been so many that I cannot list them all.
Thanks to all of you guys.

In addition, I would like to recognize:

Fred Adams, Fred Adams Paving, NC, Morrisville, NC (Chairman of ICPI Construction Committee)
Kirk Bauer, Riverside, CA
Chuck Beckman, Site Technologies, Roswell, GA
Joel Hedrick, Meadowood Pavers, Plymouth, MN
Barry Hofer, Interlocking Stone Systems, Doylestown, PA
Keith Lotthammer, Glacial Ridge, Inc., Willmar, MN
Pat McCrindle, McCrindle Paver Systems, Collingswood, NJ
Pat O'Hara, Syrstone (retired), McGraw, NY
Keith Waylen, East Penn Pavement Co., Slatington, PA

I wish to thank Bob Cramer, Field Services Manager and Deta Halilovic, Media Manager for extensive editing, proofing and persistence.
A special note of thanks to Ted Corvey of Pine Hall Brick, whose gentle but firm urging was the only reason I completed this book.

INDUSTRY VENDORS

Bernd Andreas, Rampf Molds Industries
Marshall Brown, ACM Chemistries
Paul Croushore, Bayer Corporation
Vern Dueck, Pacific Precast Products
John Dziekan, Pristine Products
Glenn Kerr, Kerr Lighting
Lee Martin, Zenith Equipment
Mike Mueller, TEKA North America, Inc.
Helga Piro, SF Concrete
Martin Probst, PROBST GmbH
Fred Schultz, CTI

CONCRETE PAVER MANUFACTURES who continue to wholly support PAVE TECH, its full line of products and services. It is important to note that almost every concrete and clay paver manufacturer in North America supports PAVE TECH by selling or using one or more of its services or products. Listed are those who deserve our special recognition.

Air Vol Block, San Luis Obispo, CA
Borgert Products, St. Joseph, MN
Cambridge Pavingstone, Lyndhurst, NJ
Capitol Concrete, Topeka, KS
Cindercrete Products Ltd, Regina, SK
Grupo Carmelo, Sabana Seca, PR
Ideal Concrete Block, Westford, MA
Kirchner Block and Brick, Bridgeton, MO
Lehi Block, Lehi, UT
Yorkton Concrete Products Ltd, Yorkton, SK

ASSOCIATION (These people also provided extensive and valuable edits)

Ted Corvey, Chairman of BIA Paving Committee, e-mail: *tcorvey@pinehallbrick.com*
Leroy Danforth, BIA Jr., EIT Manager, Technical Publications, e-mail: *danforth@bia.org*
David Smith, ICPI Technical Director, e-mail: *DSmith@bostromdc.com*
Brian Trimble, Regional Director, International Masonry Institute, e-mail: *btrimble@imiweb.com*

INTERNATIONAL

Sebastian Mueller-Kleeb, Sebastian Mueller AG., Rickenbach, Switzerland
Toshiro Ueda, Ueda Co., LTD, Sapporo Japan

SPECIFIERS

Many thanks and great appreciation for the landscape architects, designers and engineers who have helped make PAVE TECH and its products the industry benchmark.

Contents

1

Segmental Pavement Overview

SEGMENTAL PAVEMENTS

Segmental paving is the construction of flexible pavement with clay and concrete pavers. This type of paving system must contain a minimum of the following:

Main components

- Compacted aggregate base
- Sand bedding layer
- Edge restraint
- Pavers (wearing course)

Optional components

- Sub-base layer
- Geotextile
- Sub-surface drainage
- Stabilized sub-grade

Elements of a segmental pavement (Figure 1-1)

Segmental paving relies heavily on technology learned from the asphalt industry for base design and construction. Segmental paving originated in Europe for flexible industrial, municipal and commercial uses.

The system was introduced to North America for commercial work, but with the addition of colored pigments the market quickly moved to a residential focus. Now, as the residential market matures, we see a re-focus on commercial and industrial projects.

As more contractors are specializing in segmental pavement construction, there is a pressing need for a trained labor force. Trained labor is necessary for paver contractors to fill the growing demand. Our national associations have only partially filled the contractors needs for labor training. It is the desire of PAVE TECH to help fill that crucial gap in industry support with this Vocational Skills Training Manual.

Sealed paver patio (Figure 1-2)

Why PAVE TECH?

The first reason is that our industry associations failed to do this on their own. Also, the growth of our industry is threatened with decreased quality and standardization of work.

Mechanical paver laying at the Port of Oakland (Figure 1-3)

In 1985, Stephen Jones formed PAVE TECH, INC. as a residential paver installation company. His experience in those years opened his eyes to the reality of the contracting side of the industry. His first job, a 4800 sf. driveway, sidewalk, patio, pool deck proved to be one of his most complicated ever. This was the first and last time he ever used treated lumber edging. After being totally dissatisfied with the use of treated lumber for edge restraint, he began to use what has been called a "concrete toe", placed after the pavers had been laid. In Steve's case this consisted of ready-mix concrete and rebar troweled into place. But even this had difficulty staying put when vehicles traveled near the edge. This difficult and labor intensive method caused Steve to search the market for alternatives. He found none suitable.

In the fall of 1986, Steve's thoughts turned to the design of a new product specifically designed for paver installations. By spring of 1987, Steve had his first pieces of what was then called " PAVE TECH Edging". That first year, he sold about 50,000 ft. of edging. After the experiences, trials and tribulations of that first year, they relocated manufacturing to a Chicago based plastics company. It was also the time for many changes to the size and shape of what would become known as "PAVE EDGE". 1988 was also the year that "PAVE EDGE Flexible" became available to make radius work easier. The second year also saw sales increase to more than 500,000 ft. Today, PAVE TECH has companies that manufacture PAVE EDGE in the United States, Canada, Japan, New Zealand, Australia, Europe and South Africa.

PAVE TECH stands for quality and service to the industry with active involvement in the ICPI (Interlocking Concrete Pavement Institute), the BIA (Brick Industry Association) and the NCMA (National Concrete Masonry Association). PAVE TECH's many contributions to the paver industry include: PAVE

Steve Jones has presented training seminars throughout the country (Figure 1-4)

TECH's publication of the "PAVE TECH EDGE" newsletter, on site demonstrations, seminars, training and product videos.

To date, PAVE TECH has been involved in over 11 national TV shows including "PBS", "Hometime", "The Learning Channel", "The Discovery Channel" and "Home Savvy". Even though PAVE TECH stopped contracting work in 1990, PAVE TECH continues to be involved in research and education and works with its contractor customers on large and unique installations around the country.

Contractor training has always been a priority for PAVE TECH.
(Figure 1-5)

PAVE TECH introduced their chemical treatment products business for pavers in 1989. PAVE CHEM has evolved into our industry's most comprehensive line of products and support materials. This program was developed especially for professional contractor use. Along with chemicals, PAVE CHEM has also compiled the widest selection of masonry adhesives for every application along with SANDLOCK, an organic joint sand stabilizer additive. All products have been field tested before sales to customers are allowed.

The biggest addition has been the German manufactured PROBST line of hardscape installation tools and equipment. PROBST produces the finest material handling equipment available to manufacturers and distributors. They also provide the finest trade tools for contractors. This product line is a perfect fit for both companies because a philosophy of "NO COMPROMISE" is the common bond between PROBST and PAVE TECH.

As PAVE TECH grows and travels the world, we look to add to our products in support of our contractors and manufacturers. Stephen Jones is always available for suggestions, especially from contractors in the field.

PAVE TECH on the road again.
"The Road Warriors"
(Figure 1-6)

2
Concrete Pavers Defined

MANUFACTURING CONCRETE PAVERS

Concrete pavers are manufactured using all natural, environmentally friendly raw materials, including various gradations of washed sand and crushed stone, Portland cement, pigments for coloring, and water. Through computerized weighing, mixing and transferring, these materials are delivered to the paver manufacturing machine. It is deposited into various shaped steel molds, vibrated and compacted. The concrete pavers are layered directly on production pallets and cured. Next, they are ready for quality control inspection, packaging and shipment.

Interlocking concrete pavers are produced on state-of-the-art, dedicated, mechanized manufacturing equipment. Manufacturers must meet or exceed the American Standard of Testing and Materials (ASTM) C 936 standard specification for solid concrete interlocking paving units.

Concrete paver standards

The following concrete paver properties are required in ASTM C 936, the industry product standard. A concrete paver is defined as a unit that has a maximum surface area of 100.75 square inches (0.065 sq. m.). The overall length of the unit divided by its thickness does not exceed four. The length divided by the thickness is called the aspect ratio. Many current shapes exceed the dimensions set in the ASTM C 936. Check with your supplier for oversized shapes appropriate for flexible paving systems.

Concrete paver machine
(Figure 2-1)

ASTM C 936 has various tests and standards that manufacturers must meet. They include:

- Compressive strength
- Absorption
- Freeze and thaw durability
- Abrasive resistance
- Permissible variation in dimensions

Compressive strength

Compressive strength is the crushing force applied to a paver's surface that causes it to break. A concrete paver's comprehensive strength should average 8,000 psi. As a comparison, the average poured concrete in municipal sidewalks ranges from 3,000 to 4,000 psi. Concrete pavers have substantially higher strength than ordinary poured concrete.

Concrete paver plant
(Figure 2-2)

Most pavers can break when bent. They are not as easily crushed because concrete is strong when compressed, and not as strong when bent or split. When pavers do break they usually break from being bent or split by a sharp, concentrated force.

Absorption

When immersed in water, a concrete pavers' average absorption of water should be less than 5 percent (<5.0 %). In the manufacturing process, concrete pavers become a dense product.

Freeze and thaw cycles

Concrete pavers when subjected to 50 freeze/thaw cycles shall have no breakage, with no loss in dry weight greater than 1 percent (<1.0 %).

Abrasive resistance

The average thickness loss should not exceed 1/8 in. (3 mm). This is determined from the abrasion index. The abrasion index is the ratio of the cold water absorption and the compressive strength. Lower values indicate a denser unit which is more resistant to abrasion. A sandblast test for concrete pavers is an alternative for the abrasion index.

Freeze thaw conditions
(Figure 2-3)

Banding station
(Figure 2-4)

Concrete paver heights 10 cm, 8 cm, 6 cm.
(Figure 2-5)

Permissible ASTM variations in dimensions

- **Length and width variance:** Paver unit length and width should not differ any more than +/- 1/16 in. (+/- 1.6 mm).

- **Paver unit height:** Height should not differ more than +/- ⅛ in. (+/- 3.2 mm).

- **Size:** Concrete pavers units should not be greater than:

    6½ in. (165 mm) in width

    9½ in. (240 mm) in length

- **Aspect Ratio:**
    Many pavers offered today exceed ASTM dimensions that limit the surface area to 100.75 in² and a length to thickness ratio of 4 or less. Pavers that exceed ASTM dimensions standards should require special consideration when installed in any flexible paving system. They are not normally recommended for vehicular traffic.

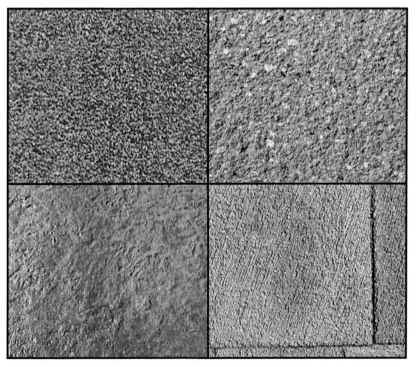

Concrete paver textures
(Figure 2-6)

Pavers are available in a variety of color and shapes.
(Figure 2-7)

Color, texture and shape

Manufacturers create concrete pavers in many colors. Most pigments are synthetic iron oxides. Numerous surface finishes can be made on concrete pavers.

Manufacturers also produce pavers in shapes ranging from squares and rectangles to circle and fan patterns. The various shapes, colors and textures allow the specifier and end user unlimited creativity in choosing pavement design patterns.

3
Clay Pavers Defined

MANUFACTURING CLAY PAVERS

A majority of clay pavers are made by the extrusion method. Clay or shale is ground to a fine consistency and conveyed to a pugmill where it is mixed with water. It then passes through a brick machine's vacuum chamber where air is removed, causing the clay particles to fit closer together. Clay is forced through the head of the brick machine like toothpaste coming out of its tube. This extrusion creates a solid column of clay called a "log". The column is then cut with the wires into individual paver units through wire banks and then loaded onto kiln cars. After about 25 hours in a dryer at 400°F, the brick pavers enter a natural gas, coal, or sawdust fired kiln for about 30 hours. The most popular type of kiln is a "tunnel" kiln that gradually raises the temperature of the pavers to over 2000°F before exiting the other end of the kiln.

Moulded clay pavers

The moulded clay process is similar to the firing process with some differences in the production process. Moulded pavers are produced from wet clay that is pressed into a wood-box mould. The excess clay is struck off from the top of the mould box. Sand or water is used to coat the wood moulds to allow the clay pavers to fall out of the mould easily when the box is turned over. The pavers are then dried and fired. The nature of a wood mould process purposefully makes each unit slightly different from each other where the extruded process produces units with more consistent dimensions.

Finished pavers are then unloaded simultaneously from several kiln cars by hand or with robots. This helps assure a blend of subtle shade differences that gives brick its warmth and character. The bricks are then packaged in "cubes" and delivered.

Clay brick extruder forming a column of clay (Figure 3-1)

Clay paver plant kiln and kiln cars (Figure 3-2)

Clay paver properties

To the untrained eye, all pavers look alike. There is a difference between clay and concrete pavers. They are both available in different sizes, shapes and specifications. After time, you will know exactly which pavers will and will not meet your requirements for a particular installation. This is important because paver properties are factors that dictate the beauty, performance and quality of a job.

Clay paver properties include:

- Color
- Durability
- Size
- Texture and Edge Treatments

Robotic stacking of clay pavers (Figure 3-3)

PAVE TECH ©

Color

The color of fired clay depends upon its chemical composition, the firing temperatures and the method of firing control. Of all the oxides commonly found in clays, iron probably has the greatest effect on color. Clay containing iron in any form burns red, making it the most common color available. Kaolin clay and fire clay burn to a light-white to buff range. Mixing various clays together, offers many different color ranges including: browns, buffs, pinks and grays. Regardless of the color selection, the vitrification process makes clay paver color permanent.

Clay pavers may also achieve a variety of colors through the addition of sands or ceramic coatings to the paver surface. The sands also facilitate extraction from their moulds are often colored to achieve a variety of looks on moulded pavers.

Flashing

Flashing takes place when excess fuel is placed into the kiln causing a fuel rich fire. The fire consumes oxygen to burn the extra fuel. It pulls the oxygen out of the raw material causing the paver to permanently change color. Flashed or full range color was common to pavers from the old beehive kilns of the late 1800's and remains the most requested color blend today.

Master color samples

Manufacturers use a master sample of each style to grade production pavers in order to maintain high quality standards. Pavers are made in large quantities, one color at a time. These are called lots or runs. Each paver run is generally tested for color, size, and durability. Although each run of brick may vary slightly in color, the variation will not dramatically change the appearance of a paver in the finished pavement. This makes additions or expansions easier.

For some larger projects, master color samples are prepared to show the blend of color in different pavers. These panels are normally set up on or near the construction site.
(Figure 3-4)

ASTM C902 STANDARDS	
Classification SX (severe weather)	
Compressive strength (min.)	8000 psi
Water absorption	8% max
Saturation coefficient	.78
Type I (severe abrasion)	.11
Application PS, PX	
(dimensional tolerances-plus or minus 8 in. dim.) ¼ in., ⅛ in.	

ASTM C 902 Standards
(Figure 3-5)

Durability - ASTM

Clay paver durability is ensured through conformity to either of two ASTM standards: ASTM C 902 Pedestrian and Light Paving Brick and ASTM C 1272 Heavy Vehicular Paving Brick. Each standard has minimum requirements of compressive strength, water absorption, abrasion resistance, dimensional tolerances, and in the case of C 1272, breaking load.

Clay paver durability results from incipient fusion where intense heat (2000°F) brings the clay or shale particles to the melting or fusion point forming a permanent bond between the particles. Since compressive strength and absorption in clay pavers also are related to the firing temperatures, these properties, together with the saturation coefficient, predict paver durability for most clay pavers. Because of differences in raw materials and firing methods, a single value of compressive strength or absorption will not indicate paver durability for some types of clay pavers. As a result, ASTM C902 allows alternate testing measures such as freeze thaw testing to substitute for some standards.

Freeze thaw

Another measure of durability is the Canadian test for F-T resistance to deicing salts. This test method was developed for concrete pavers (CSA-A231.2). The maximum allowable loss from a paver after 50 freeze thaw cycles is 500 grams. The average loss in the test was only 7.8 grams. Clay pavers typically can pass this test.

Ice and ice melt on clay pavers.
(Figure 3-6)

Clay paver thickness for flexible pavements can vary from 1 in. to 3 in. (Figure 3-7)

Size Figure 3-7

Clay pavers are available in a wide range of sizes. The most common thickness is 2¼ in. (56 mm) with the length being twice the width. *For example:* (4 in. x 8 in.) (100 mm x 200 mm) and modular size for mortared applications (7⅝ in. x 3⅝ in.) (193 mm x 90 mm). Other common sizes are thicker pavers for heavy vehicular applications (2⅝ in. minimum) (66 mm). For light threshold applications, thinner units (1¼ in.)(32 mm) are commonly available.

Texture and edge treatments

Clay pavers come in a variety of textures including wire cut, sand finish, square edge, beveled edge and antique edge. Textures produced are generally by the type of manufacturing method.

Wire cut texture Figure 3-8

The most common surface formed by the wires that cut the individual pavers in the extruded manufacturing process.

Square edge Figure 3-9

Often associated with the term "wire cut", square edge pavers are the most common clay pavers and as their name implies, the paver edges are sharp. These pavers are generally the most economic in terms of price but slight chipping occurs on the edges over time. This happens under normal usage as the paver edges rub against one another even under pedestrian traffic. These pavers generally do not feature spacers, so achieving proper joint width is important. The tendency is to lay the pavers too tightly.

To many, slight edge chipping forms a natural patina and gives the paver a weathered, old world look that is part of the charm of genuine clay pavers.

> *Note: Do not use a standard plate compactor with square edge pavers. Use a rubber roller attachment or a rubber pad to reduce chippage.*

Chamfer/Beveled edge Figure 3-10

Chamfered pavers are a popular choice as many manufacturers (extruded and moulded) are offering these pavers in addition to square edge pavers. Chamfered pavers enhance the bond line/joint line that pavers create and allow for the individual shape of each paver to be more pronounced.

WIRE CUT is easily recognized by the small clay particles dragged by the cutting wires leaving small grooves in the surface. These grooves add character as well as skid resistance to the paver surface.
(Figure 3-8)

Square edge clay paver
(Figure 3-9)

Chamfered / beveled clay paver
(Figure 3-10)

 PAVE TECH ©

Chapter 3 • Clay Pavers Defined

13

Repressed clay paver
(Figure 3-11)

Moulded or sand finished clay paver
(Figure 3-12)

Rubber roller for plate compactor on clay pavers.
(Figure 3-13)

Note: Beveled edges help reduce pavers from chipping over time and allows for the use of a steel plate compactor. During compaction it is still recommended to use a rubber pad or specialty roller attachment.

Repressed pavers Figure 3-11

Repressed pavers are typically extruded pavers that are placed, by hand, in a press machine to create bevels on the edges. They are generally more expensive because of this second pressing operation. The bevels are precise and even all the way around, providing a very consistent appearance.

Moulded & sand finish pavers Figure 3-12

Moulded pavers, by their vary nature, offer the original "antique" look and have the advantage of modern firing standards to insure long term durability. These pavers are commonly referred to as "sand" finish pavers because the pavers feature a sand coating on one surface and a "struck" side on the other as described above. Either the sand or the struck side can be used for the paver surface, although the sand finish can wear over time.

Some of the differences between moulded clay pavers and extruded clay pavers are:

- Moulded pavers offer a classic "pre-20th" century look. Moulded pavers look hand-crafted as opposed to mass produced by machine.

- The slight non-uniformities of moulded pavers, which give them their unique appearance, may require more adjustment of individual pavers during installation than their extruded counterparts. However, most moulded pavers meet the same dimensional tolerance standards as the majority of extruded pavers (ASTM C-902 Application PS).

- The sand embedded into the surface of moulded pavers is susceptible to damage from high-pressure cleaning techniques that may be acceptable on some, but not all extruded pavers. This is important to remember when mortar or grout is used and gets onto these types of paver surfaces. Installers should always obtain cleaning and other installation guidelines from the brick manufacturer, no matter what type of clay paver is used.

Used antique pavers were very popular for a while, commanding high prices. But used pavers may not be durable due to poor firing practices of the past. As a result, use caution when considering such products.

Spacers (Lugs, nibs and spacer bars) Figure 3-14

Pavers made with spacers may be used in mortarless applications. Spacers should be no more than ⅛" (3 mm) in size. They ensure a space between the pavers and provide a uniform gap for jointing sand. Spacers should not touch each other when installed. Spacers also keep the paver edges from touching. This reduces the amount of chipping during compaction and use.

Always include the spacers when measuring the specified dimensions of the paver and when laying out a paving pattern. When measuring pavers with spacers include the spacer dimensions on 1 side only.

Shrinkage for clay pavers

During the process of firing, clay pavers shrink. The amount of shrinkage depends on various factors including, but not limited to:

- Moisture content
- Type of clay
- Temperature of the kiln
- Time in the kiln

Spacers on clay pavers
(Figure 3-14)

Clay shrinks during drying and firing, so there may be unevenness in the paver size. The principal problem for the installer is not that they shrink, but the amount of resulting shrinkage that varies with each clay paver. Shrinking increases with higher temperature firing. Higher temperatures produce the more popular darker shades. Consequently, when a wide range of color is desired (flashed color), some variation always occurs between the sizes of the dark and light units. To obtain products of uniform size, manufacturers attempt to control factors contributing to shrinkage. Because of variations in raw materials and temperature variations inside the kilns, absolute uniformity is impossible. Consequently, specifications for clay paver production include permissible size variations.

*When calculating size, only add one side of the paver lugs. Example: (8 in. + ⅛ in.) **NOT** (8 in. + ⅛ in. + ⅛ in.)*

Adjusting for clay paver shrinkage is an installation issue. The installer should measure a sample of pavers delivered to the jobsite. Laying modules should be based on the largest size paver.

4

Paver Shapes and Patterns

PAVER SHAPES

Concrete and clay pavers are available in dozens of sizes and shapes. They combine to create endless designs. Often times a project will incorporate two or more shapes in the same area to create unique designs. Shown here are just a few of the basic shapes and patterns that can be installed. Some shapes need edge units to provide a clean line between the pavement and adjacent ground cover. These units are straight on one side and fit into the shapes of the other pavers on the other sides. Many projects incorporate two or more shapes in the same area to create a design. The design of a paver project can be done by a landscape architect, homeowner or contractor. Master color samples and design samples help to show the customer what the pavers will look like when completed.

Brick shaped rectangular paver
(Figure 4-1)

Master color samples.
(Figure 4-2)

Pavers come in a variety of shapes and sizes.
(Figure 4-3)

8cm Industrial pavers
(Figure 4-4)

6cm Residential concrete pavers
being used for a patio
(Figure 4-5)

SELECT THE RIGHT PAVER FOR THE APPLICATION
Minimum thicknesses for applications:

• **Residential walks and driveways:**

Concrete 2 + in. (6 cm)

Clay 1½ + in. - (5 - 6 cm)

• **Commercial and heavy traffic:**

Concrete 3 + in. (8 cm)

Clay 2¾ in. (7 cm)

Some designs are only available in 7 to 8 cm. heights.
Thicker pavers are usually specified for a project by an engineer.
Learn about new styles and patterns through your local distributor.

Combining different paver shapes,
colors and patterns
(Figure 4-6)

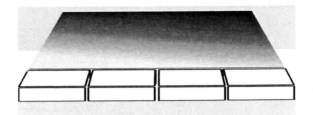

SAILOR course:
Pavers are laid end to end to outline the pavement.
(Figure 4-7)

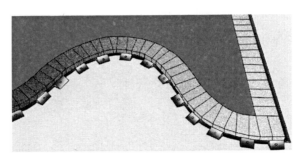

SOLDIER course:
Pavers are laid long side by side to accent the edge of the pavement. Tight curves usually require cutting off some of the Soldier Course to minimize gaping.
(Figure 4-8)

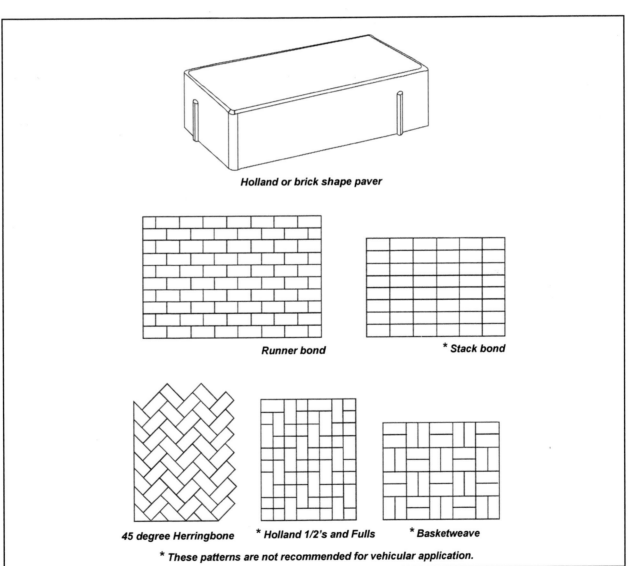

Holland or brick shape paver

Runner bond

* Stack bond

45 degree Herringbone

* Holland 1/2's and Fulls

* Basketweave

* These patterns are not recommended for vehicular application.

Holland stone or brick shaped paver with laying patterns
(Figure 4-9)

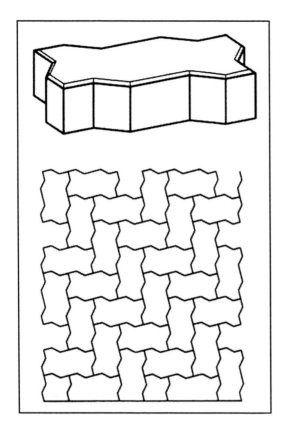

Multiweave
(Figure 4-10)

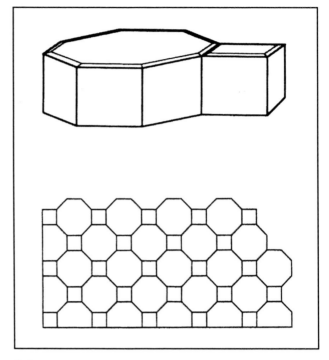

Octo
(Figure 4-11)

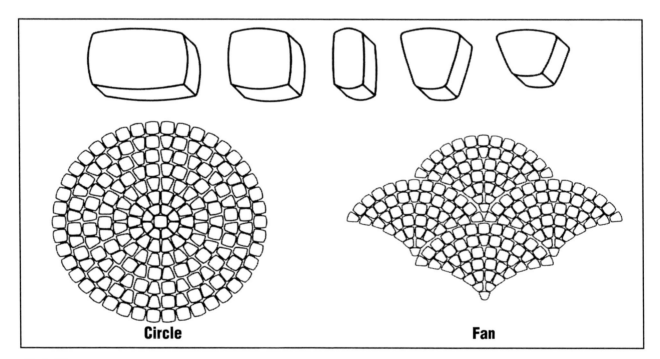

Circle

Fan

Circle / Fan
(Figure 4-12)

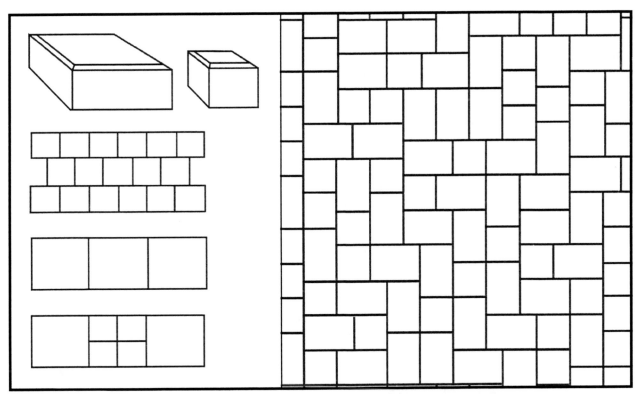

Patterned Multi-size
(Figure 4-13)

2 Piece Random Cobble
(Figure 4-14)

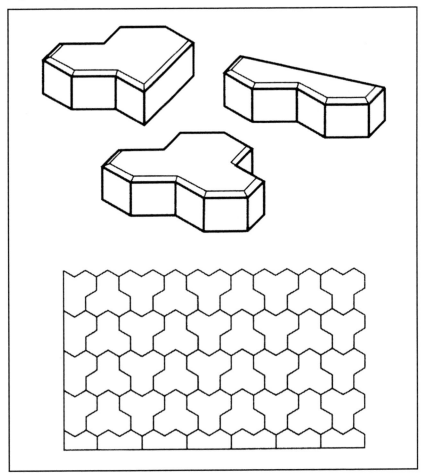

Delta
(Figure 4-15)

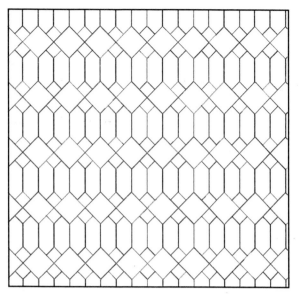

Symmetry
(Figure 4-16)

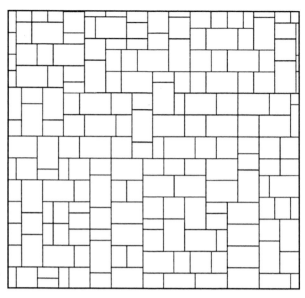

4 Piece Random Cobble
(Figure 4-17)

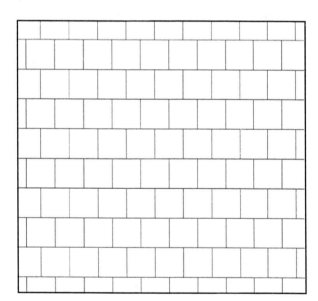

Large Cobble Running Bond
(Figure 4-18)

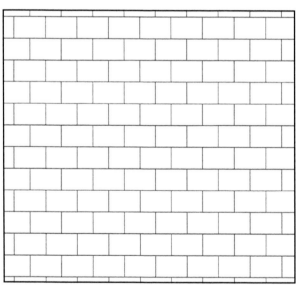

Small Cobble Running Bond
(Figure 4-19)

Some Paver patterns and shapes supplied by E. P. Henry and are copyrighted.

5

Using the Right Tools

HANDTOOLS

Before starting a paving project, it is important to be equipped with the right tools to make your job easier.

Bandcutter *(Figure 5-1)*
Bandcutters are used to safely cut metal or plastic bands on pavers.

(Figure 5-1)

Broom, Push *(Figure 5-2)*
The broom should be sturdy and have stiff bristles, such as this unit that is made especially for pavers.

(Figure 5-2)

Baserake *(Figure 5-3)*
The base rake levels and smooths the base, and evenly spreads joint sand on pavers for drying.

(Figure 5-3)

Broom, Small Hand *(Figure 5-4)*
Used for cleaning areas prior to using adhesive or mortar.

(Figure 5-4)

Chalk Lines, red or blue *(Figures 5-5, 5-6 & 5-7)*
Used to mark bond lines in bedding sand or on pavers when cutting or on base for setting edging.

(Figure 5-5)

(Figure 5-6)

(Figure 5-7)

Chisels *(Figures 5-8, 5-9)*
Assorted sizes are used for chiseling of concrete and stone. They are available in long, short, narrow and wide sizes.

(Figure 5-8)

(Figure 5-9)

Crow Bar *(Figures 5-10, 5-11)*

Used for pulling edge restraint spikes. Grind the end for easier spike pulling.

(Figure 5-10)   *(Figure 5-11)*

Diamond Blade *(Figure 5-12)*

Diamond blades are used for cutting pavers.

(Figure 5-12)

Dust Control Vacuum *(Figure 5-13)*

A vacuum system for saws. Collects dust from cutting. Use at dust sensitive sites.

(Figure 5-13)

Edge Restraint, Industrial, Rigid and Flexible

Use to maintain perimeter interlock.
(Figures 5-14, 5-15 & 5-16)

(Figure 5-14)

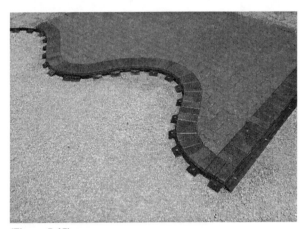

(Figure 5-15)

(Figure 5-16)

Grade Stakes

Wood stakes can be marked directly on. Steel grade stakes with nail holes are best.
(Figures 5-17, 5-18, 5-19, 5-20)

(Figure 5-17)

(Figure 5-18)

(Figure 5-19)

(Figure 5-20)

Extractor *(Figure 5-21)*

The Paver Extractor is used to remove a single defective paver. Much easier and faster then using screw drivers. Eliminates damage to surrounding pavers.

(Figure 5-21)

Flat Bar *(Figure 5-22)*

Use for miscellaneous prying of wood. Also good for prying pavers in tight areas.

(Figure 5-22)

Grinders, 7"— 9" Metal Grinder and Diamond Cup Grinder

Metal abrasive *(Figure 5-23)*

For sharpening buckets and shovels.

(Figure 5-23)

Diamond Cup *(Figure 5-24)*

For concrete.

(Figure 5-24)

PAVE TECH ©

Hacksaw *(Figure 5-25)*
Used to cut plastic edging. Suggested blade: 24 - 32 teeth/ inch.

(Figure 5-25)

HAMMERS

Hammer, 3 lb *(Figure 5-26)*

(Figure 5-26)

Hammer, Brick
Use a brick hammer to chip paver edges.
(Figure 5-27)

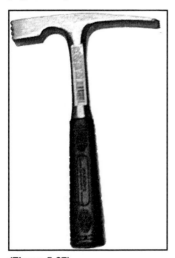

(Figure 5-27)

Hammer, Claw
Use a claw hammer to nail forms.
(Figure 5-28)

(Figure 5-28)

Hammer, Spike *(Figure 5-29)*
Specialty hammer for driving edging spikes.

(Figure 5-29)

Hammer, Setting *(Figure 5-30)*
Used for setting natural stone or adjusting large pavers.

(Figure 5-30)

Hammer, Persuader *(Figure 5-31)*
Rubber head hammer for pavers.

(Figure 5-31)

Hammer, Sledge (10 lb.) *(Figure 5-32)*
Used for breaking sections of concrete or occasionally tamping base in tight areas.

(Figure 5-32)

Levels, 2' and 4' *(Figure 5-33)*
Used to check for level grade.

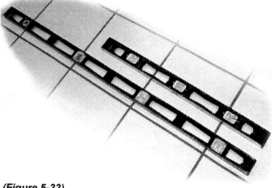

(Figure 5-33)

Trowel Margin or Putty Knife 2" *(Figure 5-34)*
Used to clean up disturbed bedding sand. Handy for getting between pavers and to smooth sand when a paver is extracted.

(Figure 5-34)

Marking Tools, Markers *(Figure 5-35)*
Markers can be soapstone, permanent markers or carpenters pencils.

(Figure 5-35)

Marking Tools, PaverScribe *(Figure 5-36)*
Used to transfer cutting angles and depth for cutting pavers to fit when **not** using a soldier course.

(Figure 5-36)

Marking Tools, QuickDraw *(Figure 5-37, 5-38)*
The QuickDraw marks the pavers for cutting when using a soldier course.

(Figure 5-37)

(Figure 5-38)

Marking Tools, 4 x 45° (Figure 5-39)

(Figure 5-39)

Marking Tools, Flex Marker (Figure 5-40)

(Figure 5-40)

Trowel, Mason (Figures 5-41, 5-42)

A Mason trowel is used to smooth bedding sand when marks have been made in it and to remove excess sand.

(Figure 5-41)

(Figure 5-42)

Nails, Cut (Figure 5-43)

Concrete square nails. Used to attach wood forms to concrete.

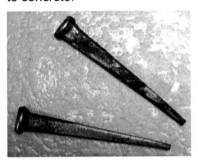

(Figure 5-43)

Nails, Duplex (Figure 5-44)

A double headed nail used on forms for easy disassembly.

(Figure 5-44)

String Lines (Figures 5-45, 5-46)

Used to check bond lines and grades.

(Figure 5-45)

(Figure 5-46)

String Collars *(Figures 5-47, 5-48)*
Used with Steel grade stakes to adjust height.

(Figure 5-47)   *(Figure 5-48)*

PaverAdjuster *(Figure 5-49)*
Used to adjust and straighten bond lines.

(Figure 5-49)

PaverPaws *(Figure 5-50)*
Used to lay pavers that are to large to lay by a single hand.

(Figure 5-50)

Pliers, Channel Lock *(Figure 5-51)*

(Figure 5-51)

PaverCart *(Figure 5-52)*
Used to transport banded and unbanded pavers on a job site. Can also be equipped with a tray for carrying equipment. PaverCart removes bands from pallet.

(Figure 5-52)

PaverCart, Conveyor Section *(Figure 5-53)*
Use a conveyor section to mover pavers on a PaverCart through doors and gates.

(Figure 5-53)

Third wheel helps with balance. *(Figure 5-54)*

(Figure 5-54)

PAVE TECH ©

Pinch Bar *(Figure 5-55)*
Use to bust up concrete, level objects.

(Figure 5-55)

Rake, steel tooth garden
Use for raking out base or knocking down compacted high spots. *(Figure 5-56)*

Rake, leaf
Used for raking sand and gravel from grass. *(Figure 5-57)*

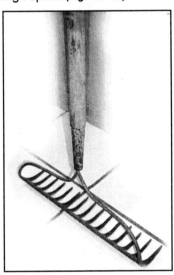

(Figure 5-56)

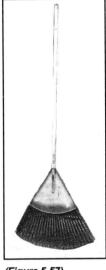

(Figure 5-57)

Square, Builders (small) *(Figures 5-58, 5-59)*
The small builders square is used to mark pavers for cutting. The large builders square is used to : "square" small areas of pavers.

(Figure 5-58)

(Figure 5-59)

Large Builders Square 12" x 18" *(Figure 5-60)*

(Figure 5-60)

Square, folding *(Figure 5-61)*
Used to create 90 and 45 degree angles and to set long square reference chalk lines.

(Figure 5-61)

Sand Spreaders *(Figure 5-62)*
Used to spread sand on large paver projects.

(Figure 5-62)

Saw, Masonry Gas *(Figure 5-63)*
Contractor equipment for fast cutting of pavers.

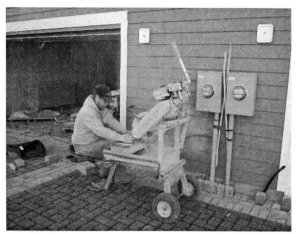
(Figure 5-63)

Saw, Masonry *(Figure 5-64)*
Electrics for smaller jobs.

(Figure 5-64)

Saw, chop *(Figure 5-65)*
Used for small repairs or homeowner use.

(Figure 5-65)

Saw, Hand held gas *(Figure 5-66)*
Used for asphalt and concrete pavement cutting. Occasionally used for cutting pavers but not recommended for that use for safety reasons.

(Figure 5-66)

Screed Guides
A ¾" steel water pipe or 1" steel electrical conduit pipe placed on the base to allow screeding of 1" bedding sand. Do not use PVC or other plastic, it must be able to bridge small imperfections. Square tubing is expensive and difficult to lay flat on the base.
(Figure 5-67)

Screeds *(Figure 5-68)*
A wooden or aluminum length of material. For wood a 2" x 6" or 2" x 8" board can be used. For aluminum a 1" x 4" or 1" x 5" board works well. Having different sizes from 3-10 ft is helpful.

(Figure 5-68)

PAVE TECH©

Screeds, hand *(Figure 5-69)*
Use a SandPull to fill screed rail marks and free float difficult areas. Floats on the sand.

(Figure 5-69)

Screeds, Large mechanical
Large mechanical screeds such as the SandMax I and II are aluminum or steel. Some allow telescoping to adjust for screed width. Most are pulled by a machine such as a loader, others are pulled by two men.

SandMax I *(Figure 5-70)*

(Figure 5-70)

SandMax I *(Figure 5-71)*

(Figure 5-71)

SandMax II *(Figure 5-72)*

(Figure 5-72)

Screeds, Standing *(Figure 5-73)*
Standing screeds such as a SandPull Pro are made from aluminum and allow the operator stand and screed bedding material by using legs for strength.

(Figure 5-73)

Shovels, Flat and Round *(Figures 5-74, 5-75, 5-76)*
Two types should be available on the job site. Flat and Round. Welded backs on the shovels are stronger but make for a heavier shovel. Long handles are easier on the lower back. Wood handles give fewer blisters than fiberglass or plastic handles, but need to be treated with linseed oil once a year.

(Figure 5-74)   *(Figure 5-75)*   *(Figure 5-76)*

SlabGrabber *(Figure 5-77)*
Used to grab large slabs or retaining wall blocks. Use on pavers to place large pavers by hand.

(Figure 5-77)

Splitter *(Figure 5-78)*
For use on DIY, tumbled pavers or large industrial type applications.

(Figure 5-78)

Transit, Optical and Laser *(Figures 5-79, 5-80)*

Optical (Figure 5-79)   *Laser (Figure 5-80)*

Sweepers, gas powered *(Figure 5-81)*
Used for cleaning and sweeping off joint sand and debris from pavers or a job site.

(Figure 5-81)

Tape Measure
(Figure 5-82)
Quality tape measures are invaluable. Everyone on the crew should be wearing one during excavation and base preparation.

(Figure 5-82)

Wheel barrow *(Figure 5-83)*
Use a wheelbarrow with one or two wheel design. It should have high quality bearings.

(Figure 5-83)

Wood Hand Saw *(Figure 5-84)*
Used to cut lumber and tree roots at a job site.

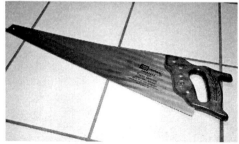

(Figure 5-84)

SAFETY EQUIPMENT

Dust Masks *(Figure 5-85)*
OSHA requires a minimum of 2 head bands and a bendable steel nose bridge. Should always be used in dust situations.

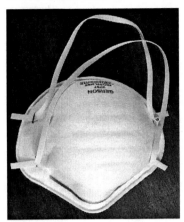

(Figure 5-85)

Finger Tape *(Figure 5-86)*
Finger tape is used to protect fingers when using rough product or in wet conditions.

(Figure 5-86)

Gloves *(Figure 5-87)*
Hand protection, shown gloves are vinyl coated polyester.

(Figure 5-87)

Hearing Protection *(Figure 5-88)*
Hearing protection is required around any noise producing activities such as sawing, compacting, or while other machine operation is going on.

(Figure 5-88)

KNEE PROTECTION

KneePad *(Figure 5-89)*
Used to protect workers knees from moisture, cement, cold and sharp rocks.

KneeSeat *(Figure 5-90)*
A KneeSeat is an alternative to traditional knee pads offering greater comfort and less stress on the knees when kneeling.

KneePad   *KneeSeat*

(Figure 5-89)   *(Figure 5-90)*

Safety Glasses
(Figure 5-91)
A good pair of safety glasses must protect the front and sides of the eyes. They should not be too dark of a tint and should be worn whenever impact striking, cutting or splitting.

(Figure 5-91)

 PAVE TECH ©   **Chapter 5 • Using the Right Tools**   **35**

MATERIAL HANDLING EQUIPMENT

Clamp, Band (Figure 5-92)
Designed to lift 1 or 2 straps or bands of pavers at one time.

(Figure 5-92)

Clamp, Cube (Figure 5-93)
Designed to lift whole strapped cubes of pavers.

(Figure 5-93)

Pallet Wagon (Figure 5-94)
Allows for pallet movement on a job site by hand.

(Figure 5-94)

Skid steer (Figure 95)
Used for excavating, base grading and material handling.

(Figure 5-95)

Slab Vacuum, MammothMite 110 (Figure 5-96)
Allows for 2 man lifting and placement of large concrete or stone slabs on a job site.

(Figure 5-96)

Mechanical Laying of pavers (Figure 5-97)
Able to lay full layers of pavers from a cube (approx. 7 -14 sq. ft.) at one time.

(Figure 5-97)

Mechanical Laying of pavers *(Figure 5-98)*

(Figure 5-98)

Laying Machine *(Figure 5-99)*

Equipment designed purposefully to mechanically lay pavers at a fast rate.

(Figure 5-99)

COMPACTION EQUIPMENT

Use **Hand Tamper**, (Pounder) to compact small difficult to get at areas. Use the **Chisel Point** to remove unwanted concrete. *(Figures 5-100, 5-101)*

(Hand Tamper 5-100)

(Chisel Point 5-101)

Jumping Jack Tamper *(Figure 5-102)*

Used for tight areas and corners also for foundations and utility trench compaction.

(Figure 5-102)

Forward Plate Compactor *(Figure 5-103)*

5 hp gas, 4000 lb force.

(Figure 5-103)

Rubber Mat
(Figure 5-104)
Bolts to plate for protection of square edge clay pavers and textured concrete pavers.

Rubber Roller
(Figure 5-105)
Attachment for clay pavers, textured surface concrete pavers and large slab pavers.

(Figure 5-104)

(Figure 5-105)

Forward Plate Compactor Diesel *(Figure 5-106)*
4.9 hp, 5000 lb force

(Figure 5-106)

Reversible Plate Rammer *(Figure 5-107)*
Depending on model, usually 10,000-13,000 lbs of force.
Travels both forward and in reverse.

(Figure 5-107)

Double Drum Walk Behind *(Figure 5-108)*
Vibratory roller compactor

(Figure 5-108)

Double Drum Ride-on *(Figure 5-109)*
Vibratory roller compactor

(Figure 5-109)

PAVE TECH ©

6
Utility Locating

KNOW WHERE TO DIG

Before beginning any excavation, call the local utility companies (including water, gas, electric, phone, cable TV). The first rule regarding underground utility location: **Never assume where existing utilities might be**. Have utilities located and marked before beginning any excavation. *"Call before you dig."* The number to call is usually in the front of local phone books. Normally, you can call just one number and that organization will coordinate all the local utilities to mark the site. If not, call each utility individually. This service is provided at no cost. Plan for approximately 3-5 working days for all the utilities to mark the locations. Remember to write down your call confirmation number.

Mark all utilities around the job site. (Figure 6-1)

Warning tape is sometimes placed one lift above underground utilities. (Figure 6-2)

Warning tape for gas line. Sometimes it is laid directly with utility line which then gives no warning. (Figure 6-3)

Gas line. (Figure 6-4)

Utilities to call: *(if there is not a local utilities "one call")*

- *Electric*
- Natural gas
- Telephone
- Cable television

Tell the local utility companies:

- What you will be doing.
- When you will be working.
- Where the excavation will take place on the property.
- How deep you will be digging.

Have all underground utilities marked. If the markings do not look right, or if you have questions about them, call the utility again and discuss it. Utilities are allowed about a 3' leeway on either side of their markings so be cautious. To avoid damaging underground lines, it is important to locate the utilities by digging by hand. Sometimes utility company's will lay warning tape one lift above the utility (Figure 6-2). It is still necessary to properly locate the utility by hand.

Sprinkler systems:

Underground sprinkler systems are easily damaged. It may be necessary to relocate parts of a sprinkler system. It is a good idea to have an agreement in writing with the owner. This should include who is responsible for relocating the sprinkler system. It should also cover who is liable for the cost of any repairs if damage should occur during construction. As a precaution, locate the water shut off valve cap and remove sprinkler heads until the installation is complete. If cutting any sprinkler lines or removing heads, either fold over the line and tape it together or block the inlet to stop dirt from getting into and clogging heads once reattached.

Layout, Elevation and Drainage

LAYOUT

Layout is transferring the design from paper to the job site. Adjustments often need to be made to the design due to an awkward corner, curve, or the natural grade of the site. Layout is the time to address these considerations and adjust for them.

Familiarize yourself with site conditions. Consider access, structures in or next to the paving area and jobsite material handling. The pavement design can be supplied by the customer or others. It can also be done by a physical layout on the job, then a drawing done on site by the contractor on graph paper.

(Figure 7-2)

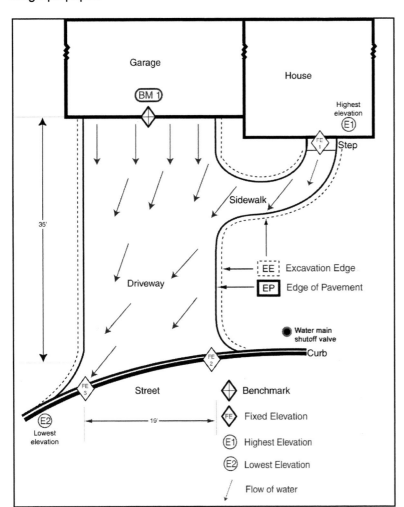

Garage

House

BM 1

Highest elevation

E1

FE 1 — Step

Sidewalk

35'

EE — Excavation Edge

EP — Edge of Pavement

Driveway

● Water main shutoff valve

Curb

FE 2

FE 3

E2 — Lowest elevation

Street

19'

⬥ Benchmark

FE — Fixed Elevation

E1 — Highest Elevation

E2 — Lowest Elevation

Flow of water

Stakes and paint lines mark the layout of the new driveway.
(Figure 7-3)

The driveway in our example will be constructed to slope to one side allowing good drainage.
(Figure 7-1)

Is this the way your stringlines look?
(Figure 7-4)

Pull string tight
(Figure 7-5)

Tying a stringline
Simple but important

Stringlines should be used throughout a jobsite. They are constantly being tied and untied. To tie a stringline should therefore be a quick, easy, simple task.

1. Pull the line tight to the stake.
2. Loop the line around the stake.
3. Loop it around two more times
4. Hold the two ends tight.
5. Bring line across top of the loops on the stake. The line will hold.
6. To release the line, pull the loose end.

Loop around stake once. (Hold taught with one hand and wrap with other.) (Figure 7-6)

Loop around stake twice.
(Figure 7-7)

Loop around stake three times.
(Figure 7-8)

Lift line. Push stringline
(Figure 7-9)

Let go. To release just pull loose end.
(Figure 7-10)

Set up collars once for a job
(Figure 7-11)

Fast take down and re-set
(Figure 7-12)

Try using steel grade stakes and string collars to speed up the layout and elevation checking process throughout the job.
(Figure 7-13)

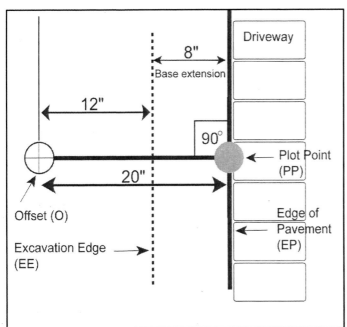

Edge Elements
(Figure 7-14)

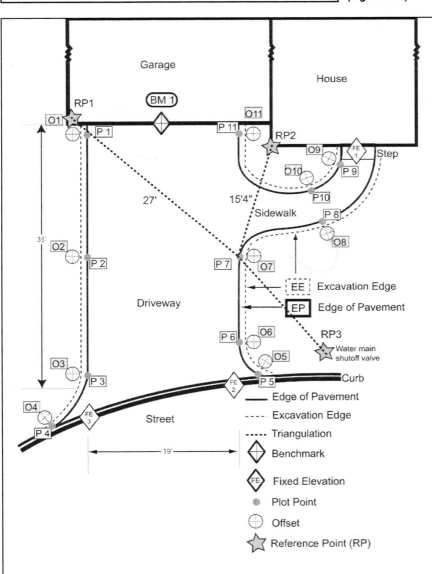

Elements
(Figure 7-15)

Elements of layout

- Base Extension
- Benchmark
- Drainage
- Edge of Pavement
- Elevation
- Excavation Edge
- Fixed Elevations
- Grade
- Offset
- Plot Point
- Reference Lines
- Reference Point
- Surface Square Footage
- Triangulation

Creating Center Perpendicular Reference Line

The most important reference line on the job is the *center perpendicular reference line*. Bond lines, edging and other reference lines are determined from it. The center reference line normally is parallel to the length of the pavement and perpendicular to the structure the pavement will meet.

To create the center reference line, here is a field technique that is very useful. (Figure 7-17, illustrates the technique).

- Determine the center of the main body of pavement. In our example, the driveway is abutting the garage slab. The driveway will be 19 feet wide. The center would be at 9-1/2 feet from the edge.

- On the garage slab measure 8 feet from the center point (A) to each side of it. Mark these two points.
  Note: This measurement can be any number but it is important to use the same measurement on each side of the center point.

- Make two measurements of equal length to the point where they cross each other. We will use 20 ft. in our example.

- Mark where the two measurements arcs intersect. This is the center point (B).

- Snap a chalk line using the center mark (A) of the garage slab and the point where the two lines meet. Extend this line to the curb. Mark on the curb. This line is the center point (B) center perpendicular reference line.

Mark the center perpendicular reference line with either stringlines or preferably with a chalk line.
(Figure 7-16)

Creating Parallel Reference Line *(Figure 7-16)*

It is important to create a parallel reference line to start laying pavers. This helps by making sure your first laid pavers are square and true. The 1st parallel reference line could be created in two ways. (See example).

1. Measure equally at a 90° from the front of the garage slab down to near the curb on either side of the driveway. C-D + E-F. Then snap a chalkline across D-F.
   or
2. Create a 90° line off of your center perpendicular reference line.

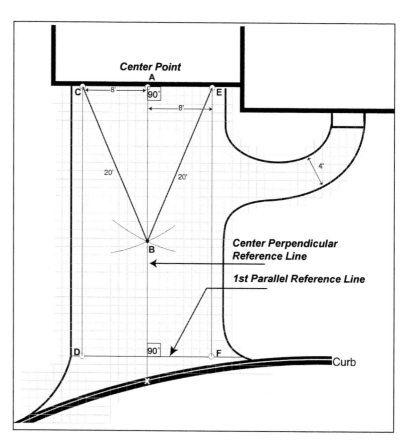

Center Perpendicular Reference Line
Point A to Point B is perpendicular to
the garage slab.
(Figure 7-17)

Base Thickness

Flexible segmental pavements vary in required compacted base depth.

The minimum for any segmental pavement is 4 inches of base, providing that best site conditions exist and the pavement is only for pedestrian use.

The thickness of the base is determined by application, expected load, soil conditions and climate. ICPI recommends the following <u>minimum</u> thicknesses:

NATIVE SOIL	Sand	Soil	Clay
EXPECTED USE	BASE THICKNESS		
Sidewalk or pation	4"	5-6"	6-8"
Driveway	6"	8"	10-12"
Municipal Pedestrian	8"	10-12"	12-18"

Base Thickness Chart (Figure 7-18)

Creating Offsets from Plot Points *(Figure 7-19)*

Offsets

Offsets (O) are created using the plot points (PP). During excavation the original Plot Points will be continually disturbed, so it is necessary to create Offset stakes . These stakes will also be used to measure elevation and base depth. The Offset (O) is located outside of the excavation area. It is normally placed 12" from the Excavation Edge (EE) at a 90 degree angle to the Edge of Pavement (EP) with each corresponding Plot Point. Offsets are used to quickly relocate plot points by measuring into the excavation 20". So, for our example (Figure 7-19) 12"+ 8"(base extension), our offsets are 20" from our Plot Points (PP).

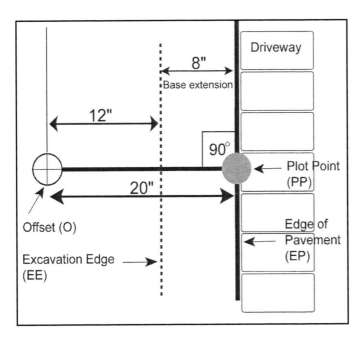

Creating Offsets (Figure 7-19)

Terms for layout

Edge of Pavement (EP)

The edge of pavement is where the proposed paver surface ends.

Excavation Edge (EE)

This mark is crucial because it defines the boundaries of the excavation.

Benchmark and Fixed Elevations   *(Figure 7-20)*

Benchmark (BM) is the reference elevation for all other elevation measurements. When selecting a Benchmark, look for something permanent that will not be disturbed during construction.

For our illustration example we have chosen the benchmark (BM) as the garage slab that abuts the proposed pavement. The paver installer selects the benchmark on most residential projects such as patios or driveways. On large projects, a benchmark is given on the site plan.

Fixed Elevations (FE) are the elevations of fixed points that the new pavement will meet. These will not be moved or changed during the construction process. Examples of fixed elevations are curbs, steps, slabs and sidewalks.

We will use 3 Fixed Elevations for our site. The house step and two points on the curb. These points will not be disturbed during construction and can be used to determine elevation. (See figure 7-20).

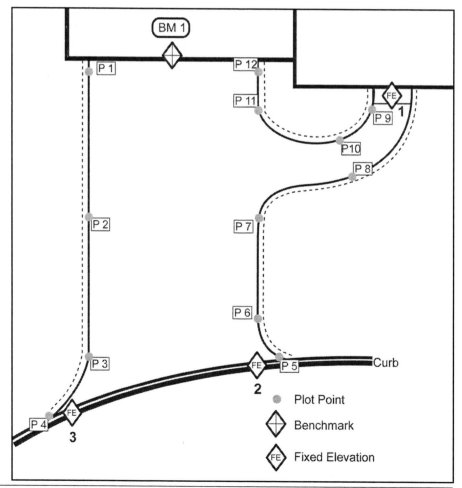

***Benchmarks Fixed Elevation
Plot Points
(Figure 7-20)***

			ELEVATION CHART		
Location	Existing Grade	BOE Bottom of Excavation	1st Lift Base Compacted +3"	2nd Lift Base +3"	3rd Final Base Grade
BM					
FE 1					
FE 2					
FE 3					
P 1					
P 2					
P 3					
P 4					
P 5					
P 6					
P 7					
SIDE WALK P 8					
P 9					
P 10					
P 11					
P 12					

(Figure 7-21)

Plot Points (PP) are placed on the Edge of the Pavement (EP). They are typically placed where the proposed pavement meets with a structure, existing surface or changes direction or elevation. They are used to chart elevations. The more contour in a pavement design or elevation, the more Plot Points. Don't overdo the Plot Points. If the layout is being done on site, mark the points on a drawing. If the contractor is following a drawn design of the project, transfer the plot points from the plan to the site using marker paint. Later on, Plot Points will also be used to run string lines to check finished grade.

Elevations

Accurate soil subgrade elevations are critical for uniform layering of the aggregate base material. As the paver installer, you may or may not be responsible for doing elevations. On large projects, the project specifications should clearly state who is responsible for setting elevations. It could be the site engineer, surveyor, or whomever is building the base.

Marking elevations

Write the elevation for each finished grade elevation on the Offset stake or set a string collar. With steel stakes, use tape and permanent markers to mark lines and elevations or use string collars and a chart. Write the elevation for each plot point. Normally set string collars at offsets 12" above finished base elevation. (See figures 7-22 and 7-23).

> *Example: Finished elevations, for the top of the pavement, are written on the stakes, and they are shown as a plus or minus from the benchmark on the plans.*

Verify that all the finished excavated elevations are within 1/4" (6 mm) of the plan. If the elevation is too high, keep digging.

If the elevation is too low, more aggregate base material will be needed. This not only effects base construction but also will end up costing more for labor, equipment and material. Do not over dig!

(See Elevation Chart, pg. 47).

Steel grade stake. Showing elevations on duct tape with a marker. (Figure 7-22)

Wood grade stake. Marks are made directly on the wood. (Figure 7-23)

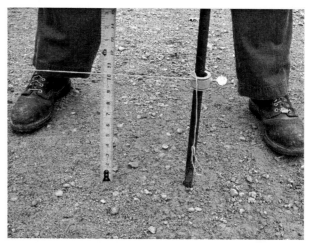

Strong collars allow fast take down and re-set (Figure 7-24)

Stakes mark the Offset Points
(Figure 7-25)

Checking base depth at bench mark.
(Figure 7-26)

Triangulation Chart

	RP 1	RP2	RP3
P1			
P2			
P3			
P4			
P5			
P6			
P7	27'	15' 4"	20'
P8			
P9			
P10			
P11			
O1			
O2			
O3			
O4			
O5			
O6			
O7			
O8			
O9			
O10			
O11			

Triangulation Chart
(Figure 7-27)

Triangulation *(Figure 7-28)*

Triangulation is the method to re-identify a point (either Plot Point or Offset). Triangulation is used to confirm Plot Points and Offset locations during excavation and base construction. Excavation of a site may remove or disturb these. Through triangulation, these points can be reinstalled.

Reference Points (RP)

To start, Reference Points are selected. These Reference Points are permanent points on the site that will not change during construction. Select three points to use, mark them to create a triangulation chart. Sometimes you will only need to measure 2RPs to establish your PPs. Make a chart to track your triangulation numbers. (See Figure 7-27).

Measure the distance from the Reference Point to each Plot Point. Note the measurement on the chart. Continue until all of the Plot Points are positioned by triangulation. Do the same with the Offsets.

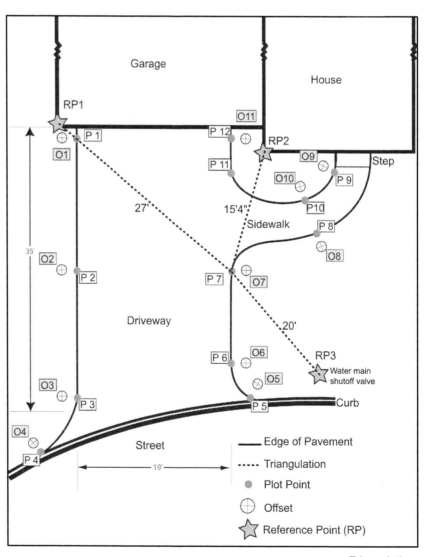

Triangulation
(Figure 7-28)

Drainage Plan

Planning ahead for surface water drainage is very important. A pre-site visit will help to determine the natural grade of the pavement and the required adjustments to achieve good drainage. A drainage plan prior to excavation determines where the water will flow on the site. At the very least, a 2% grade away from the house or buildings should be maintained. It is important to understand and map the flow of water.

The longer (distance) water is kept on the paver surface, the more volume is accumulated. This includes an increase in water velocity that creates a destructive force. From the beginning of any pavement job it is critical to factor the drainage. By creating a plan, you will know how the grade must be set to achieve optimum water flow.

To control water on the pavement, contractors make use of crowns, swales and sheeting profiles. Crowns are made by raising the center of the pavement to shed water off both edges. Swales are created by lowering a section of pavement creating a linear valley. These can be dangerous because they channel the water and increase the flow area. Sheeting is the practice of creating a flat surface of a consistent grade across a large area of pavement. This moves the water off the pavement with the least destructive effect.

Often times the drainage plan for a project will require a crown, swale, or sheeting profile of the pavement.

Crown in the pavement is a slight lift to the profile in the middle of the pavement that moves surface water to each side and off the pavement.

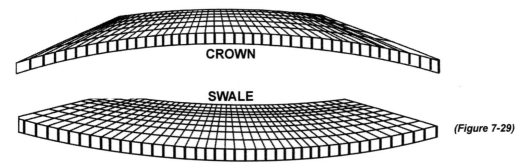

(Figure 7-29)

Swale is a slight 'dip' in the pavement which channels water on the pavement.

Sheeting is when the pavement surface remains flat yet still has a grade. This allows for water to move off the surface without building a large volume of concentrated run-off.

Determine the highest and lowest points of the site. Ideally, a newly constructed pavement on a level lot would crown in the middle and slope at

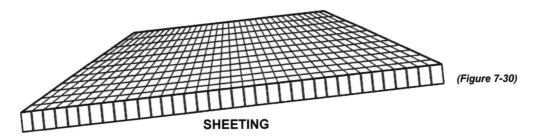

(Figure 7-30)

least 2% to the curb and sides. Often times natural slope prevents that sort of grade. A grade is illustrated in Figure 7-31 that moves the flow of water across the pavement to the lawn and curb. The highest point is located at the corner of the house E1 and the lowest on the lower left curb when entering the driveway E2. Constructing the base and achieving the final grade should reflect the optimum flow of water according to the drainage plan.

It is important in residential paver construction to keep the pavers higher than surrounding land-scape to allow for fast drainage into the grass or soft areas. The landscape slows the speed of the water and allows it to percolate into the soil.

On our example (Figure 7-31) E1 is the highest elevation and E2 is the lowest elevation.

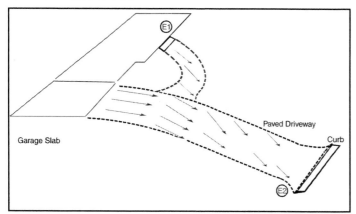

Plan the flow of water on the site with a drainage plan.
(Figure 7-31)

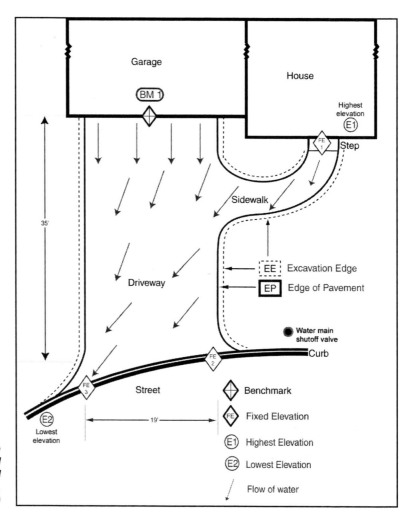

The driveway in our example will be constructed to slope to one side and down to the street allowing good drainage.
(Figure 7-32)

8

Excavation and Materials Calculation

EXCAVATION

Excavation is the digging and removal of materials from the pavement site. Excavation begins after layout is complete.

Prior to the actual excavation, accurate estimates are made to know the weight and volume of the materials being removed. These figures determine how much material will be removed and how much material will be needed to construct the pavement.

From the layout, we determined the shape and surface square footage of the project. To make use of formulas, we also need to know the depth and types of material being removed.

Removing native soils.
(Figure 8-1)

The dimensions may be thicker in colder parts of North America, or in soft clay soils, or those that are frequently saturated.

Our example will have a 4 inch base for the sidewalk and an 8 inch base for the driveway.

NATIVE SOIL	Sand	Soil	Clay
EXPECTED USE	BASE THICKNESS		
Sidewalk or patio	4"	5-6"	6-8"
Driveway	6"	8"	10-12"
Municipal Pedestrian	8"	10-12"	12-18"

Base Thickness Chart (Figure 8-2)

Base Extension (BE)

The base extension extends beyond the edge of the pavement. The base for any flexible pavement system must be wider than the actual paved surface. This spreads the load and stress from the pavement to the base. The base extends past the pavement edge a distance equal to the depth of the base. For driveways with an 8 inch base the base extension is 8 inch. For sidewalks with a 4 inch base, the base extension is 4 inch.

Bedding Sand

1 inch of sand is loosely screeded prior to laying pavers. The bedding sand will compact to approximately 3/4"+/- 1/8" when the pavement is completed.

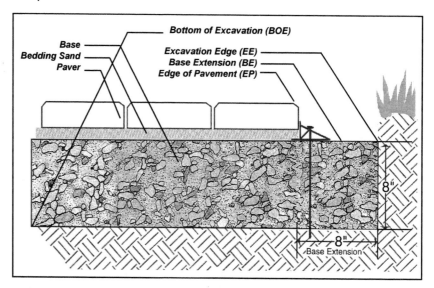

Base Extension Illustration (Figure 8-3)

Aggregate base extends beyond Edge of Pavement to create a shoulder. (Figure 8-4)

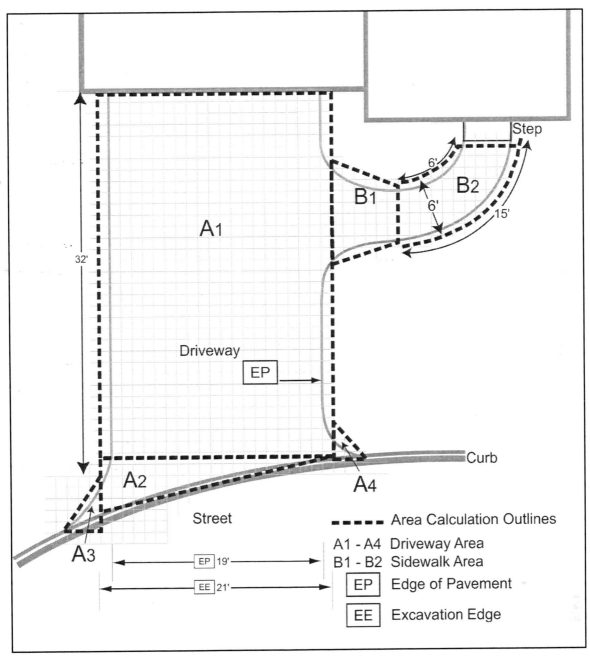

Surface square footage
(Figure 8-5)

Surface square footage

Determine the surface square footage of the driveway and sidewalk areas separately.

"A1" Driveway

To simplify the math, divide the driveway layout into rectangles and triangles.

For rectangles the formula for square feet is:

> Length (ft) X Width (ft) = Total square feet
> Example: Driveway (A 1): 32' x 21' = 672 sf
> EE is actually 20.33', but for a safety margin, we will use 21'.

The formula to determine the area of a triangle is:

> Length x Width ÷ 2 = sf
> Example: Driveway (A3)

NOTE:
For example purposes we will use all measurements in feet and inches

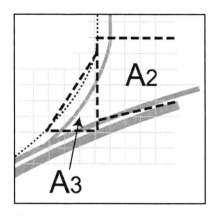

Curved areas can be best estimated by easily creating overlaying triangles.
(Figure 8-6)

Curves on a layout are considered triangles (A2, A3 and A4). Average the location of two sides of the triangle. This means cutting in one side and making a straight line between two points.

"A2"

Length = 21'
Width = 5'
21 x 5 = 105 ÷ 2 = **52½ sf.**

"A3"

Length = 3'
Width = 5'
3 x 5 = 15 ÷ 2 = **7½ sf.**

"A4"

Length = 2'
Width = 3'
2 x 3 = 6 ÷ 2= **3 sf.**

A2	52½	Triangles
A3	7½	
A4	3	
	63	**sf**

Now add Main driveway

Main driveway **A1**	672 sf
"*triangles*" **A2-A4**	+63 sf
Total	**735 sf**
Driveway	

The surface area of the driveway is 735 sf. This figure will be used to determine both the amount of removed excavated material and the amount of aggregates that will be necessary to build the base.

"B" Sidewalk

To determine the square footage of the sidewalk, divide it into geometric shapes.

B1 is the part of the sidewalk where it meets the driveway *(Figure 8-8)*.

First, add the two parallel sides together. Then, divide the total by 2. This will give the average length. Next, determine the perpendicular height (6'). Finally, multiply the average length (7') by the perpendicular height to get B1 sf.

"B1"

9 + 5 = 14' ÷ 2 = 7
7' X 6' = **42 sf**

"B2"

Curved area such as section B2 can be estimated by averaging the length of both curved sides. The average of the two sides is then multiplied by the width (6') to determine the square foot area of the section.

B2 is the curved section of the sidewalk.

6' + 15' = 21' ÷ 2 = 10½'
10½' X 6' = **63 sf**

Section B1 is where the sidewalk meets the driveway.
(Figure 8-7)

Note:
840 sf is the total surface square feet of the driveway and sidewalk. This is useful for paver and sand calculations.

Add the two sections of the sidewalk together to determine the total surface square footage.

Sidewalk square footage:

	B1	42 sf.
	B2	+ 63 sf.
Sidewalk		**105 sf.**

Driveway	**735 sf.**
Sidewalk	**+ 105 sf.**
Total sf.	**840 sf.**

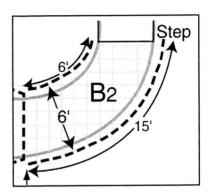

Section B2 is the curve of the sidewalk.
(Figure 8-8)

Formulas have been provided to make accurate estimates for soil and base material.

These formulas are:

- Excavation factor (Depth x sf.)

- In Place (Volume of compacted or undisturbed material)

- Swell factor (Volume of loose material)

Use these formulas to determine the weight and cubic yards of the material that need to be removed and the amount of materials needed to construct the pavement.

The square footages of the project are:

Driveway:	**735 sf**
Sidewalk:	**105 sf**

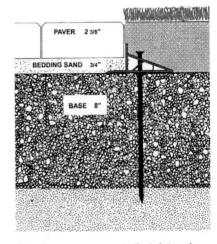

The three components that determine excavation depth are paver thickness, bedding sand and compacted base.
(Figure 8-9)

Excavation factor

Excavation factor is a formula that determines the cubic yards of material in the ground. It is a handy formula to use when the depth is known in inches and the area is known in square feet.

Excavation depth

Determining excavation depth depends on the depth of the base, bedding sand and height of pavers.

The thickness of the base is determined by usage.

- 4 – 6 in. for residential walks and patios
- 6 – 8 in. for pool decks
- 8 – 12 in. for residential driveways

The minimum for any project with segmental pavement is 4 in. of base, providing that best site conditions exist and the pavement is only for pedestrian use.

Our example will have a 4 in. base for the sidewalk and an 8 in. base for the driveway.

1 in. of bedding sand compacts to ¾ in.

Pavers are available in a wide variety of thicknesses, choose a paver suitable for the pavement application. For light vehicular traffic the minimum paver thickness recommended is 2-2¼" (50-56 mm). For example purposes we have selected a 2⅜ in. (6 cm) thick rectangular paver.

Driveway Depth

Paver	2 3/8"	*(60 mm)*
Bedding Sand	3/4"	*(20 mm)*
Base	8"	*(200 mm)*
Total	11 1/8"	*(280 mm)*

Sidewalk Depth

Paver	2 3/8"	*(60 mm)*
Bedding Sand	3/4"	*(20 mm)*
Base	4"	*(100 mm)*
Total	7 1/8"	*(180 mm)*

EXCAVATION FACTORS AND VOLUME

Depth	Cubic Yards per Square Foot
1/8"	.00038
1/4"	.00077
1/2"	.00154
1"	.00308
2"	.00617
3"	.00925
4"	.01235
5"	.01543
6"	.01852
7"	.02160
8"	.02469
9"	.02777
10"	.03086
11"	.03394
12"	.03704
18"	.05555
24"	.07407

Excavation Factor Chart
(Figure 8-10)

Driveway
5.8 C.Y. of asphalt
19.43 C.Y. of clay/gravel

Sidewalk
1.29 C.Y. of concrete
1.02 C.Y. of clay/gravel

Totals
5.8 C.Y. of asphalt
1.29 C.Y. of concrete
20.45 C.Y. of clay/gravel

The depth of excavation for the driveway in our example is 11⅛ in., the sidewalk is 7⅛ in.

Using the Excavation Factor Chart for In Place Materials
(Figure 8-10)

Since we know the depth of the excavation and the square footage we can determine the weight and volume of the materials that need to be removed. Select the factors from the excavation chart. Since not all numbers are listed in our chart it is necessary to add the factors for 11 in. and ⅛ in.

Driveway Depth	Excavation Factor
11 in.	.03394
+ ⅛ in.	+.00038
11⅛ in. =	.03432

Driveway:

Area x (times) Excavation Factor = Cubic Yards (C.Y.)

735 sf

x .03432 Excavation Factor

25.23 C.Y.

25.23 cubic yards is the total volume of the materials in the ground that will need to be excavated from the driveway. Since the excavated materials are asphalt and a clay/gravel mix, we need to know how much of each we have.

There is 2½ in. of asphalt that needs to be removed from the site. Math tell us that 2½ in. of 11⅛ or 11.125 in. is 23% of the total excavation. (2½ = 2.5 ÷ 11.125 = .225 or 23%) of the total

23% of 25.23 (total cubic yards) = 5.8 C.Y. of asphalt.

25.23 - 5.8 = 19.43 C.Y. of clay/gravel

Asphalt	5.8 cubic yards
Clay and Gravel	+19.43 cubic yards
Total	25.23 cubic yards

Sidewalk Depth	Excavation Factor
7 in.	.02160
+⅛ in.	+.00038
7⅛ in. =	.02198

Determine the cubic yards using the excavation factor:

105 sf

x .02198 Excavation Factor

2.31 C.Y.

There is 4 in. of concrete on the current sidewalk that needs to be removed. 4 in. of 7⅛ or 7.125 in. is 56% of the total sidewalk excavation.

(4 ÷ 7.125 = .56 or 56%)
.56 x 2.31 = 1.29 C.Y.

Concrete	1.29 cubic yards
Clay and Gravel	+1.02 cubic yards
Total	2.31 cubic yards

Material Weight and Swell Chart

This chart gives the weights of material, *In Place* and *Loose*. The excavation factor determined the *In Place* cubic yards. *In Place* material is dense and compacted. After it is excavated, air is added to it 'fluffing' it. This is called "Swell". Swell varies for different types of material. The material weight chart supplies us with approximate weights of materials giving a different weight for *In Place* or *Swelled*. Trucks are limited by the amount of weight they can carry, by knowing the material weight, we can determine the number of loads.

Use the Swell Factor to determine the Volume of swelled cubic yards (Figure 8-11) of material. This is the amount to be trucked to the dumpsite. Use the Material Weight for Loose cubic yards (Figure 8-11) to determine the weight.

MATERIAL WEIGHTS AND SWELL FACTORS

MATERIAL	Per C.Y. (Loose)	Per C.Y. (In Place)	Swell Factor
* Asphalt	2400	3500	1.45
Cement, portland	2450	2950	1.20
Clay, natural _____	2700	3500	1.30
* Clay and gravel, dry	2300	3100	1.34
Clay and gravel, wet	2600	3500	1.34
* Concrete	2650	3700	1.40
Concrete, wet mix	3600	3600	1.40
Earth, dry loam	2300	2850	1.25
Earth, wet loam	2750	3400	1.24
Granite	2800	4560	1.65
Gravel, 1/4 to 2 in. dry	2850	3200	1.12
Gravel, 1/4 to 2 in. wet	3200	3600	1.13
Laterite	3900	5200	1.33
Limestone, blasted	2500	4250	1.69
* Limestone, crushed	2700	4500	1.67
Limestone, marble	2700	4550	1.69
Mud, dry	2100	2550	1.21
Mud, wet	2650	3200	1.21
* Sand, dry	2750	3100	1.13
Sand, wet	3150	3600	1.14
Sandstone, shot	2700	4250	1.58
Shale, riprap	2100	2800	1.33
Slate	3600	4700	1.30
Coral, class No. 2, soft	1760	2900	1.65
Coral, class No. 1, hard	2030	2900	1.67

Material Weights & Swell Factors chart (Figure 8-11)

*Materials used in our example.

This site is still very rough and will need a lot of hand work before it is ready for geotextile and base. (Figure 8-12)

Totals of
IN PLACE to hand out
5.8 C.Y. of asphalt
1.29 C.Y. of concrete
20.45 C.Y. of clay/gravel

Cubic Yard = C.Y.

Asphalt **Volume**

5.8 C.Y. x 1.45 (swell factor) = 8.41 C.Y. swelled

Asphalt **Weight**

8.41 C.Y. swelled x 2,400 (lbs per C.Y. from chart) = 20,184 lbs
20,184 lbs ÷ 2,000 lbs = **10** tons

Concrete **Volume**
1.29 C.Y. x 1.40 (swell factor) = 1.81 C.Y. swelled

Concrete **Weight**

1.81 C.Y. swelled x 2,650 (lbs per C.Y. from chart) = 4,797 lbs
4,797 lbs ÷ 2,000 lbs = **2.4** tons

Material Weight to remove
Combined for — Asphalt: 10 tons
Trucking Recycle — Concrete: 2.4 tons
Clay: 31.5 tons

Clay/gravel mix **Volume**

20.45 C.Y x 1.34 (swell factor) = 27.4 C.Y. swelled

Clay/gravel mix **Weight**

27.4 C.Y. swelled x 2,300 (lbs per C.Y. from chart) = 63,020 lbs
63,020 lbs ÷ 2,000 lbs = **31.5** ton

Note: All numbers are rounded up.

Excavator equipment and excavator

A common excavator for constructing segmental pavements is a skid steer loader. It should have a sharp bucket that is able to dig clay and capable of lifting a 3,000 lb. cube of pavers. Properly sized equipment is important to efficiently excavate a jobsite. A small machine may not be the most efficient and a large one may be too big for the job. Proper depth of excavation can only be achieved by a skilled excavator. A good excavator understands the need for critical tolerances and the importance of digging to the correct depth and grade. He will cut cleanly along the edges and hand dig corners when necessary. A good operator maintains the equipment in good working order by replacing worn pins and bushings and keeping the bucket edges sharp.

Bobcat type loader. Use the largest
unit that will work on the site.
(Figure 8-13)

Important to keep surface of
Excavation flat, smooth and to grade
(Figure 8-15)

Grind the bucket edge to keep it sharp.
(Figure 8-14)

Ordering aggregate materials

Base material

The base material for our example is crushed limestone. The driveway will have an 8 in. base and the sidewalk will have a 4 in. base.

Using the Excavation Factor Chart (Figure 8-10), we can determine the amount of material that needs to be ordered.

Driveway

Depth   Excavation factor
8 in.   =   .02469

735 sf x .02469 (Excavation Factor) = 18 C.Y.
18 C.Y. of **In Place** crushed limestone
18 C.Y. x 4,500 lbs = 81,000 lbs (40.5 tons)

Sidewalk

Depth   Excavation factor
4 in.   =   .01235

105 sf x .01235 = 1.30 C.Y.
1.30 C.Y. of **In Place** crushed limestone
1.30 C.Y. x 4,500 lbs = 5,850 lbs (2.9 tons)

Ordering Materials

Crushed Limestone: 46 tons
Dry Sand: 4 tons

Base Material

Driveway   40.5 tons
Sidewalk   +2.9 tons
              43.4 tons
              x1.05   for over excavation
              45.6   tons of crushed limestone
                   (We will order 46 tons.)

Bedding / Jointing sand

Bedding sand is usually coarse washed sand as discussed in *Chapter 6 Sand Properties.* Regardless of application, the bedding sand should always be at 1" loose screeded or ¾" compacted for your calculating.

To determine the amount of sand, use the surface square footage of the project and the Excavation Factor chart (Figure 8-10) to determine the cubic yards of dry sand.

Bedding Sand

Total square footage of project = 840 sf
Excavation Factor for 1 in. bedding sand = .00308
840 sf x .00308 = 2.6 C.Y. of dry sand

Use *Per C.Y. Loose* from chart (Figure 8-11)
2.6 C.Y. x 2,750 lbs = 7,150 lbs
7,150 lbs ÷ 2,000 lbs = 3.6 tons of dry sand

In our example we are able to source a single type of sand for both bedding and jointing. In some areas this is not possible and you will have to plan for delivering.

Jointing Sand

Add 10% for jointing sand (10% of 3.6 tons = .36 ton)
3.6
+.36
3.96 tons of dry sand
     (We will order 4 tons of dry sand)

9
Material Logistics

As a crew person on a paver job it is important to understand the movement of expensive equipment and materials in and out of the job site. Trucks with base, sand and paver materials are usually sub-contracted to other suppliers. To have a truck wait is expensive.

Jobsite material handling involves more than finding a place to store the pavers before they are laid. The logistics of a jobsite involve the physical movement of materials and equipment in and out of the site. Planning the logistics for a paving job is essential for timely and profitable completion. A planned time line is a good idea when beginning a job. The timeline helps to determine the number of trucks needed, when materials will arrive and be used, and generally works as a plan to complete the project.

"Logistics – The acts of planning for and providing the materials, supplies and equipment required to complete a project in a timely, efficient manner".

– Jerry Dodd, Major Ret.
U.S. Military Logistics Expert

Material handling
(Figure 9-1)

Material Weight to remove:
- *Asphalt: 10 tons*
- *Concrete: 2.4 tons*
- *Soil: 31.5 tons*

Materials to bring in

These materials will be used for the job:

- *Aggregate base material:* 46 tons
- *Bedding and jointing sand:* 4 tons
- *Geotextile:* s.f. of excavation + 15% (for overlap and edges
  840 x 1.15 (15%) = 966
- *Pavers:* 8 cubes (the pavers are ordered by others)
- *Edge restraint:* 60 ft. *PAVE EDGE* RIGID, 33 ft. *PAVE EDGE* FLEX

Equipment (Large)

- *Bobcat skid steer loader*
- *Tri-axle 15 yard dump truck*

Excavated Materials

Asphalt or concrete: Check with your recycling center for dumping these materials to determine if they must be separated or can be mixed.

Aggregates and soils: These materials can be mixed and are considered clean fill. They are easier to dump.

Locate and contract site for dumping of removed materials.

Ideally, all removed site materials would be trucked to the same location. If it is necessary to truck materials to different sites, take the time to travel to each site into consideration when determining the number of trucks needed for the job and how long it takes to load each truck.

Determine amount of material to be removed.

The excavation in our example will involve removing the current asphalt pavement and concrete sidewalk. We will also excavate the old base material and soil. Calculating the amount of material being removed will determine the number of truckloads of material.

In Chapter 8 "Excavation", we determine that the excavation depth of the driveway is 11⅛ in. We will be removing 2½ in. of asphalt and 8⅝ in. of clay and old base material. The sidewalk will have an excavation depth of 7⅛ in. It consists of 4 in. of concrete and 3⅛ in. of clay and gravel.

There will be 10 tons of asphalt, 2.4 tons of concrete and 31.5 tons of soils (clay and old base) to remove from the site.

Concrete and asphalt can be recycled in most locations in the U.S.
(Figure 9-2)

Determine number of loads needed to remove site materials

Loads are determined by weight. The larger the truck, the more efficient for dumping. When ordering trucks, make sure that the on site loader can reach into and properly load the truck.

Order enough trucks so that material removal is done efficiently. Semi-end dumps can haul approximately 22 tons of material, but cannot be loaded by a bobcat. A tri-axle dump truck can haul about 15 tons of material. Load restrictions in every state may differ; check with your hauler to know the limitations.

The truck for our example is a tri-axle dump that can haul 15 tons/load. We will require 1 load of concrete/asphalt and 3 loads of soils (old base/clay) to be removed. To efficiently remove the material, consider more than one truck to do the hauling.

Determine the number of trucks that are needed, taking into consideration the following:

- *Weight of material and capacity of the truck.*
- *Time it takes to load and unload.*
- *Travel time to and from dump site, including wait time.*

After the first truck has been loaded, the excavator can start digging the clay. From our calculations, we know that we will need to have 3 truckloads to remove the clay. With the use of a time line we can determine the number of trucks needed to remove the material in a timely, cost-effective manner.

Determine the supplier for the aggregate supply.

Order the materials needed. Order pavers, edging and any other installation supplies with enough lead time from the supplier or vendor. Confirm material order and transport services at least a week ahead. Is the transport provided or does it need to be scheduled? Schedule the arrival of aggregate materials. Will the same trucks be used for removal of materials be used for delivering the aggregate base and sand for the job? If the aggregate base and sand are being delivered by others, it is important to be ready when the delivery trucks arrive. If there is a delay, call as soon as possible to delay or reschedule the deliveries.

Excavating driveway.
(Figure 9-3)

Base Material Placement

Base delivery on top
of geotextile fabric.
(Figure 9-4)

It is best if you keep all materials
stored in the street in a neat fash-
ion to reduce potential of getting
a visit from the city. (Figure 9-5)

On site storage

Create a plan where materials will be stored until needed. *Aggregate base material: Place in driveway after excavation.*

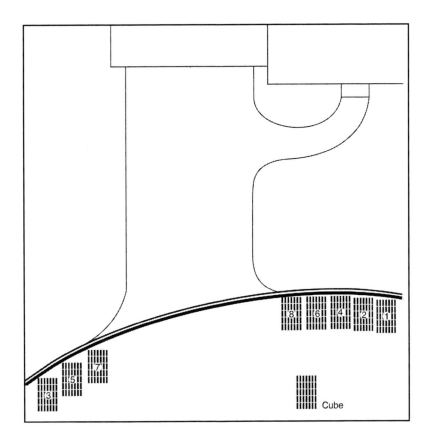

Sand Dumping
Stockpiling sand at top of driveway makes it easy to move for screeding during the job.
(Figure 9-6)

If traffic allows, place pavers in the street for easy access by paver cart.
(Figure 9-7)

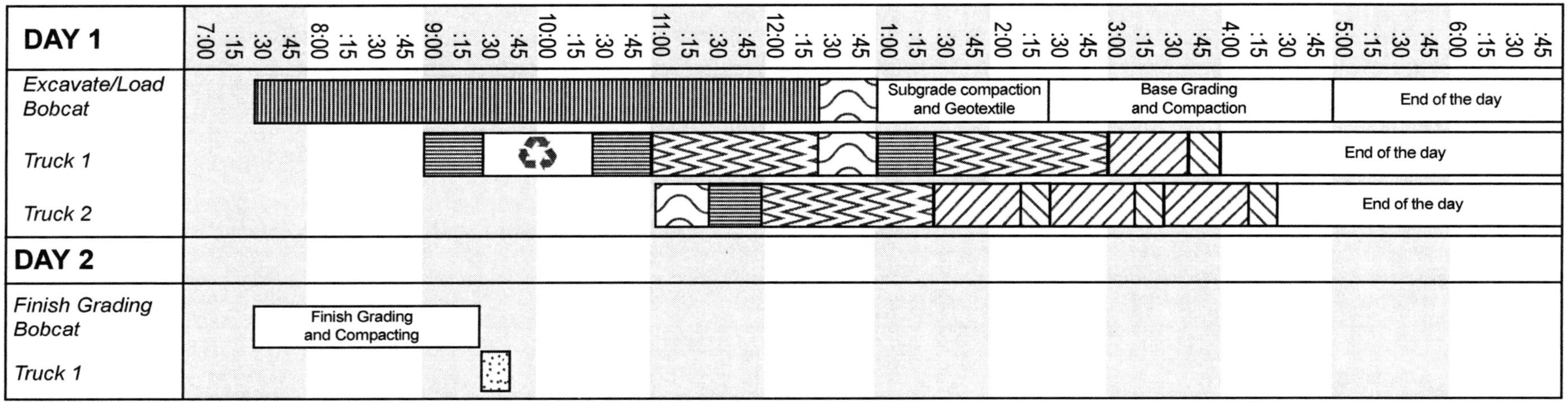

Excavation and Materials Handling TIME LINE

:45 :30 :15 7:00 :45 :30 :15 8:00 :45 :30 :15 9:00 :45 :30 :15 10:00 :45 :30 :15 11:00 :45 :30 :15 12:00 :15 :30 :45 1:00 :15 :30 :45 2:00 :15 :30 :45 3:00 :15 :30 :45 4:00 :15 :30 :45 5:00 :15 :30 :45 6:00 :15 :30 :45

DAY 1

Excavate/Load Bobcat

Truck 1

Truck 2

Subgrade compaction and Geotextile

Base Grading and Compaction

End of the day

End of the day

End of the day

DAY 2

Finish Grading Bobcat

Truck 1

Finish Grading and Compacting

Legend:

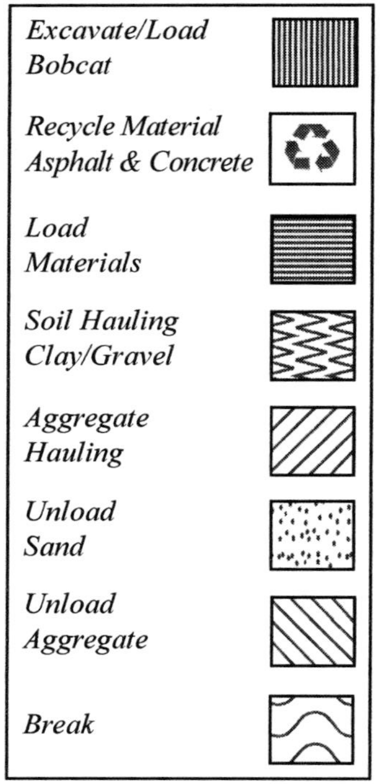

Excavate/Load Bobcat

Recycle Material Asphalt & Concrete

Load Materials

Soil Hauling Clay/Gravel

Aggregate Hauling

Unload Sand

Unload Aggregate

Break

Time it takes to load each truck: 30 min.

Travel time to and from recycle area: 1 hr

Number of truckloads to recycle: 1

Travel time to and from dump: 1hr 30 min

Number of truckloads to dump: 3

Travel time to and from aggregate pit: 45 min

Number of truckloads of base: 4

Material delivery: 15 min.

**Timeline
(Figure 9-8)**

PAVE TECH ©

Day 1

7:00 a.m.

Equipment: Bobcat   *Job:* Excavate old driveway asphalt and old sidewalk concrete.
Pile material near road.

9:00 a.m.

Equipment: Bobcat   *Job:* Load asphalt/concrete.
Equipment: Truck 1 (15 ton tri-axle dump truck)   *Job:* Load.

9:30 a.m.

Equipment: Bobcat   *Job:* Excavate old base material & clay/gravel
Equipment: Truck 1   *Job:* Leave to dump asphalt and concrete for recycling.
Return to job site.

10:30 a.m.

Equipment: Bobcat   *Job:* Load truck with old base & clay/gravel.
Equipment: Truck 1   *Job:* Load.

11:00 a.m.

Equipment: Bobcat   *Job:* Pile clay/gravel.
Equipment: Truck 1   *Job:* Leave to dump clay/gravel at clean fill site.
Return to job site after.

11:30 a.m.

Equipment: Bobcat   *Job:* Load Truck 2 with clay/gravel.
Equipment: Truck 2 (15 ton tri-axle dump truck)   *Job:* Load.

12:00 p.m.

Equipment: Bobcat   *Job:* Pile clay/gravel.
Equipment: Truck 2   *Job:* Leave to dump clay/gravel at clean fill site.
Go to Aggregate Pit after to pick-up first load of new base material.

12:30 p.m.

Equipment: Bobcat   *Job:* Lunch
Equipment: Truck 1   *Job:* Lunch

1:00 p.m.

Equipment: Bobcat   *Job:* Load Truck 1 with remaining clay/gravel.
Equipment: Compactor   *Job:* Compact subgrade and place geotextile.
Equipment: Truck 1   *Job:* Load.

1:30 p.m.

Equipment: Compactor   *Job:* Compact subgrade and place geotextile.
Equipment: Truck 1   *Job:* Leave to dump clay/gravel at clean fill site.
Equipment: Truck 2   *Job:* On route to pick-up base material.

2:15 p.m.

Equipment: Truck 2   *Job:* Deliver base material.

2:30 p.m.

Equipment: Bobcat   *Job:* Spread base material and grade.
Equipment: Compactor   *Job:* Compact base material.
Equipment: Truck 2   *Job:* Leave to pick-up more base material and
return to job site.

3:00 p.m.
    Equipment: Bobcat   *Job:* Spread base material and grade.
    Equipment: Compactor   *Job:* Compact base material.
    Equipment: Truck 1   *Job:* On route to pick-up base material.

3:15 p.m.
    Equipment: Bobcat   *Job:* Spread base material and grade.
    Equipment: Compactor   *Job:* Compact base material.
    Equipment: Truck 2   *Job:* Deliver base material.

3:30 p.m.
    Equipment: Bobcat   *Job:* Spread base material and grade.
    Equipment: Compactor   *Job:* Compact base material.
    Equipment: Truck 2   *Job:* Leave to pick-up more base material and
return to job site.

3:45 p.m.
    Equipment: Bobcat   *Job:* Spread base material and grade.
    Equipment: Compactor   *Job:* Compact base material.
    Equipment: Truck 1   *Job:* Deliver base material.

4:00 p.m.
    Equipment: Bobcat   *Job:* Spread base material and grade.
    Equipment: Compactor   *Job:* Compact base material.
    Equipment: Truck 1   *Job:* Released - done for the day.

4:15 p.m.
    Equipment: Bobcat   *Job:* Spread base material and grade.
    Equipment: Compactor   *Job:* Compact base material.
    Equipment: Truck 2   *Job:* Deliver base material and then released
for the day.

5:00 p.m.
    Equipment: Bobcat   *Job:* Pile extra base material for the night.

End of day

Day 2

7:30 a.m. - 9:30 a.m.
    Equipment: Bobcat   *Job:* Final grade.
    Equipment: Compactor   *Job:* Compact base material.

9:30 a.m.
    Equipment: Truck 1   *Job:* Deliver bedding/joint sand, then released.

Soil Types and Subgrade Compaction

SOIL TYPES

The soil subgrade is your starting point for building a segmental paver system. How much to compact the soil subgrade is determined by soil type, moisture content and the equipment used for the compaction process. It is important to have a smooth, flat surface on the compacted soil subgrade before base work begins. This surface should reflect the final grade and profile of the finished pavement. There should be no ruts or large rocks visible in the subgrade.

Geotechnical engineers divide soil into three types: clay, granular and organic. These classifications are based on the soil particles and how moisture affects the particles. Each type of soil holds water and sticks together differently.

(Figure 10-1)

Clay Soils

There are many types of clay soils. None are considered easy to work with. To work with clay for soil sub grade you need larger equipment to excavate and specialized equipment to compact. The average residential segmental paving contractor does not usually have this type of equipment. We recommend that you avoid the use of loose clay soils to adjust grades. It is just too difficult to get good compaction on loose clay. If you need to adjust grade in a clay area at the bottom of the excavation it is recommended to use aggregate base material.

Granular Soils

Sand and gravel, made up of grains as small as 0.002 in. (0.05 mm).

Organic Soils

This includes topsoil, loam, and peat, and may consist of leaves, moss, or other organic matter. Constructing a pavement over organic soils is not recommended. A Geotextile fabric is a must under these conditions.

Squeezing soil in your hand can give you an indication on soil cohesion.
(Figure 10-2)

Soil Properties

To understand soil subgrade compaction it is important to understand some of the basic industry terms.

- **Shear resistance**
- **Elasticity**
- **Cohesion**
- **Moisture content**
- **Compaction factors**

Shear Resistance

Shear resistance is how much the soil resists moving when you apply pressure, compaction, or vibration. As soil particles move by each other, there is friction, or resistance to moving. Granular soils have low shear resistance. Higher shear resistance soils need more force to compact. Clay soil is an example of high shear resistant soil.

Elasticity

This measures how well the soil can go back to its original form when a load or compacting force is taken away. Do not use soils that are high in elasticity for a paving job, road building, or construction in general. If the subgrade packs down under a heavy load, then relaxes when the load is removed, the surface ultimately breaks down.

Cohesion

Cohesion describes how much the soil particles stick to each other. Granular soil particles, like sand, are non-cohesive. Clay soil particles are cohesive; they stick to each other.

> *Example: Think of how easy it is to pick up a handful of moist garden soil, squeeze it into a ball, and have it stay like that. Compare that to doing the same thing with a handful of moist sand. The sand will fall apart.*

Moisture content

This is a measure of the amount of water in the soil or aggregate base. Moisture content is an important property to consider in getting the highest level of compaction in your job. You will need the right amount of water – enough to lubricate the particles and let them slide by each other, but not so much that the water fills spaces between soil particles and keeps them from binding together. Do not add water to clay or organic soils. It may be necessary to add some water to some soils or aggregate base to get them to compact correctly.

Compaction factor

Frequency and amplitude – are the two forces produced by compaction equipment. Frequency is how many impacts per minute. Amplitude is the theoretical height that a compactor bounces off a properly compacted surface. Compaction machines have a predetermined frequency to generate maximum force for the soil type. A machine's amplitude varies with the soil type. Amplitude changes as the soil or base becomes more compacted or dense.

Compaction equipment
(Figure 10-3)

COMPACTION

The compaction process pushes soil particles closer together forcing out air and water from the soil. This process makes the soil more dense. Unlike loose, uncompacted soil, dense soil increases support of what is on top of it, resisting movement under load.

Equipment for Compacting Soil (Figure 10-3)

Various soils and areas of the excavation require different compaction equipment to achieve proper compaction.

Hand Tamper – POUNDER (Ramming)

A hand held tool with 3 bit attachments comes with a plate attachment. Great for all types of soil and for difficult to reach edges and corners.

It comes with a 6 in. x 6 in. tamper plate. Other accessories are two different chisel points.
(Figure 10-5)

The POUNDER is a great handtool for hand compacting hard to get at areas.
(Figure 10-4)

Hand tamper for compacting areas where mechanical equipment can not reach.
(Figure 10-6)

Compactor Lift Chart

COMPACTOR	BASE	SUBGRADE	
Type	Granular Base Lift Thickness	Sand	Clay
4000 lb. Forward Plate	3 – 4 in.	2 – 6 in.	1 – 2 in.
6000 lb. Forward Plate	4 – 5 in.	2 – 8 in.	2 – 3 in.
13000 lb. Reversible Plate Rammer	6 – 8 in.	6 – 12 in.	3 – 4 in.
Jumping Jack	6 --10 in.	6 – 12 in.	4 – 6 in.
3000 lb. + Double drum walk behind roller	2 – 3 in.	2 – 4 in.	1 in.
7000 lb. + Double drum vibratory ride on roller	4 – 5 in.	4 – 6 in.	2 – 3 in.
Hand Tamper	3 – 4 in.	2 – 4 in.	2 – 4 in.

Compaction Chart
(Figure 10-7)

Forward Plate Compactors

Low amplitude, high frequency. This type has an eccentric with rotating weights, which deliver the force and travel speed. Vibration machines have a higher frequency, from 4,000 to 5,000 blows-per-minute, and a lower amplitude. These are typically 3.5 horsepower to 8 horsepower. As the eccentrics rotate, it sends a stress wave into the ground. The vibration moves the soil particles or aggregate base material. It rearranges the particles and forces the air and water out. Good for sand and gravel soil; poor for use on clay or peat.

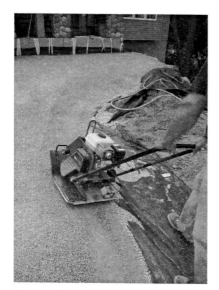

(Figure 10-8)

Plate compactor on clay soil. 1/2 in. of base material is spread to prevent the clay from sticking to bottom of plate. (Figure 10-8 & 10-9)

(Figure 10-9)

Jumping Jack (Upright Rammer)

The Jumping Jack uses a spring mechanism (high amplitude). The jumping jack is typically used on utility trench lines that cross the excavation. It has a stronger compaction force than plate compactors. The jumping jack compactor is most effective for clay and deep excavation areas. The jumping jack is essential around foundation walls, curbs, corners and where other equipment cannot reach. It can be operated in place without moving. Deep areas (12 in. or more) can be compacted at a time. Excellent for clay. Too slow for large areas.

Reversible Plate Rammer

Reversible plate rammers have a high frequency, from 3,400 to 3,700 blows per minute, and a high compaction force (11,000 to 13,000 lbs.). They are very good for clay with good production rates, but are still good for only 3 to 4 in. of clay soil at a time. Also excellent with most other types of materials.

Jumping jack compactor near a garage slab
(Figure 10-10)

Reversible plate rammer compactor
(Figure 10-11)

Vibratory Rollers Double Drum (Smooth Drum)

Vibratory rollers are fast and very efficient. Usually recommended on jobs larger than 1,500 square feet. They have a high frequency and have a compaction force of 3,000–10,000 pounds. The roller also helps smooth the base. These are typically high frequency, low amplitude machines. Available in ride on or walk behind models. Poor for clay soils, but excellent for sand and gravel type soils.

Ride on (Figure 10-12)

Walk behind (Figure 10-13)

Double drum vibratory rollers and ride on and walk behind models
(Figure 10-12,10-13)

Chapter 10 • Soil Types and Subgrade Compaction

Pad-foot roller attachment for Bobcat (Figure 10-14)

Pad-foot or Sheep's foot roller (Medium to low amplitude) *(Figure 10-14)*

This type of vibratory roller has tamping feet and compacts the soil with a kneading action; excellent for clay, but poor for sand or aggregate soils.

Static Rollers *(Figure 10-15)*

Static rollers apply only weight for compaction. Normally used on asphalt, these are not recommended for compacting soil subgrade or aggregate base material.

Preparing Soil Subgrade

When compacting clay soils, material may stick to the plate or roller. To avoid this, spread ½" of base material over the soil subgrade before compacting (See figure 10-8). Check and scrape the bottom of the plate occasionally to remove any sticking soil.

Static Rubber-tired roller (Figure 10-15)

Cement treating soils with very poor soil subgrade

When preparing the subgrade, it may be necessary to use portland cement to stabilize the soil prior to compaction. Spread portland cement, (1) 50 lb. bag per 50 sq. ft. Rake it into the soil and if necessary, add a some aggregate base material. Continue to compact the area to a smooth even surface. Stabilizing the subgrade in this manner does not replace base preparation, but helps prepare for building a base on top of it.

Utility trench areas

Utility trenches that cross the jobsite need additional compaction. These trenches can be from 3 ft. - 12 ft. deep and 6 in. - 3 ft. wide. A jumping jack tamper works best to compact these areas. The jumping jack tamper will compact deeper and delivers a more concentrated force to an area than other types of compactors. When properly compacted this area will be lower than the surrounding subgrade. Use aggregate base to bring up to grade. The harder the surface of the subgrade, the easier to achieve compaction of the base layer.

Foundations

Around foundations, curbs and utilities it is necessary to pay extra attention to the compaction of subgrade. During construction these areas are back-filled and have not settled or been properly compacted. Other areas are under overhangs of buildings where moisture is prevented from reaching the soil. Use a jumping jack tamper to compact these areas.

Weak subgrade areas

Weak subgrade areas may require a 2 layer "engineered' base design. A subgrade is built to support the sub base. This may be necessary on commercial projects or extremely poor soils. If the addition of a sub-base is required, it should be designed by a pavement engineer.

Another difficult area of subgrade is directly in front of the garage slab. This area is notorious for settling over a long time, warranting extra preparation. Directly in front of the garage slab, excavate an extra 8 inches deep and 3 feet wide gradually bring up to the level of the rest of the base depth.

Double Check the Subgrade Elevation

Check the soil subgrade elevation. Be certain the soil subgrade elevation allows for the depth of aggregate base material, pavers and bedding sand. Do not over excavate, extra costs can occur. Use a transit, levels, lasers, or stringlines from your offset stakes to check the elevation.

Complete the subgrade compaction to the best of the ability of the equipment.

Sub Surface Drainage (Figure 10-18)

When water is a concern on a project, the sub surface may require drainage. This can be done with drain tile, strip drains or filter cloth layered under the base or under the sand.

Checking subgrade elevation
(Figure 10-17)

Drain tile on patio
(Figure 10-18)

ICPI Base Training
(Figure 10-16)

11
Geotextile Fabric and Root Barriers

GEOTEXTILE FABRIC

Geotextiles are defined as any permeable textile used with foundation, soil, rock, earth or any other geotechnical engineering material as an integral part of a man-made project, structure or pavement system. Geotextiles are made of polypropylene in woven or nonwoven configurations, from yarns, fibers or slit films. They are used for drainage, filtration, stabilization and soil reinforcement applications.

There are 4 kinds of geotextiles used for base containment or stabilization:

- *Woven*
- *Non-woven*
- *Geogrid*
- *Cellular containment grid*

Woven geotextile is the preferred product for use with segmental pavements base containments and stabilization.
(Figure 11-1)

Woven
Woven geotextiles are the preferred product for segmental pavements. They are very strong and do not stretch when force is applied. Newer woven geotextile are also permeable and very cost effective. All geotextiles are available in assorted grades. Products such as Amoco's 2000 series and Mirafi 500 or 600 work well.

Geotextile fabric staked out prior to base material being laid.
(Figure 11-2)

Non-woven geotextile fabric.
(Figure 11-3)

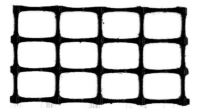

Geogrid is also used to contain base
material and stabilize subsoils.
(Figure 11-4)

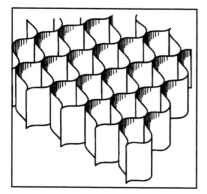

A cellular containment grid is used
to contain base material and stabilize
subsoils.
(Figure 11-5)

Geotextile is available on large rolls
(Figure 11-6)

Non-woven

Non-woven geotextiles are also permeable. They are mainly used as a filtration fabric, allowing water to pass through the material but not fine particles.

Geogrids

Geogrids have small square or rectangular shaped openings. Geotextile fabric is recommended over use of geogrid. The grid style is the least effective for segmental pavements and the most expensive.

Cellular containment grids

Cellular containment grids are three dimensional, 4 in. to 8 in. (100 mm to 200 mm) high that provide stability under loads for soils that are not cohesive, such as gravel or sand. Geogrids and cellular containment grids are not normally recommended for use when installing a segmental pavement.

Separation or Stabilization

Geotextile is used to permanently separate two distinct layers of soil materials. The classic example is where a base is to be built in poorly drained, fine-grained soil (clay or silt) and a geotextile is laid down prior to placing aggregate base material. This keeps the soft, underlying soil from working its way up into the expensive gravel and it keeps the gravel from punching down into the soft soil. The full gravel thickness remains intact and provides full support for many years.

Geotextiles enhance performance, separation, stabilization, reinforcement and drainage. One cause of pavement failure is contamination of the aggregate base, causing loss of strength. When aggregate is placed on a subgrade, the bottom layer becomes contaminated with soil. Over time, this decreases the effective aggregate thickness decreasing the support and reducing its performance and life. Geotextiles eliminate these problems.

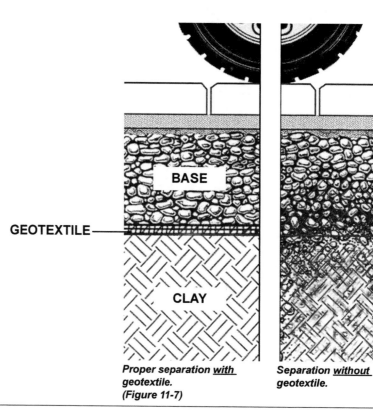

*Proper separation **with**
geotextile.*
(Figure 11-7)

*Separation **without**
geotextile.*

Using Geotextile Fabric

Prior to the installation of the base material, a decision to use geotextile fabric is made.

Geotextile fabrics are placed on top of the soil sub grade just prior to installation of the base. This gives the best support for the pavement for a number of reasons. It keeps the soil subgrade and aggregate base material separate. The geotextile fabric envelopes the base material by wrapping up the sides of the excavation. During base compaction it restricts the movement of the base material, preventing base migration during compaction. Geotextile helps increase the load capacity of the base by transferring load and prevents rutting and depressions in the pavement due to base deformation.

When to use geotextile fabric
Geotextile fabric is commonly used when the soil subgrade is primarily composed of clay or organic matter. Clay is an expansive soil and will erode the bottom of the base as it expands and contracts with the freeze-thaw cycles.

Installing geotextile fabric
Place the geotextile fabric on top of the compacted soil subgrade.

Geotextile comes on rolls of varying length. When joining sections overlap the sections of fabric a minimum 18 to 24in.

Run the edges of the fabric up the sides of the excavation and structures. Lay the aggregate base material. When geotextile is next to a concrete slab or building, the excess material is trimmed with a knife; then, with a hand held torch after the pavement is completed.

When installing geotextile, run it fully up the sides of the excavation; when the job is completed it can be cut back with a propane torch or knife.
(Figure 11-8)

Overlap geotextile at least 18 inches.
(Figure 11-9)

Spreading Base on Geotextile. It is very important that when spreading base, you do not disturb the fabric. Always try to drive over already spread base and work forward. This helps tension the fabric. Do not turn with a skidsteer until after you get some compaction. (Figure 11-10)

ROOT BARRIERS

Root barriers are a relatively new invention. They are designed to stop tree root growth into the pavement layers. They force the horizontal roots to go deeper before allowing them to move horizontally. Use of root barriers is just common sense with the high cost of pavements. It also reduces maintenance and trip hazards and, of course, unsightly lifting. Once roots have been redirected under the aggregate base into the native soil sugrade, they do not try to surface again.

Use a pruning saw to cut tree roots. Careless hacking away at a tree root can cause vascular collapse and trauma to any tree.
(Figure 11-11)

Deep Root Illustration
(Figure 11-12)

These roots show the redirection of the roots when they encountered the Root Barrier.
(Figure 11-13)

The old brick pavers surrounding this tree are performing well. Had root barriers been available when it was installed, it would last for decades more.
(Figure 11-14)

12

Base Construction and Compaction

COMPACTION

A good base or foundation is critical to a segmental pavement system. This begins in the compaction of the base. Compaction is the process of pushing aggregate particles closer together making the base material more dense. Unlike loose, non-compacted materials, a properly compacted base supports loads without rutting or moving.

Base materials

Aggregate base material is usually aggregate such as ¾ in. minus crushed stone. Different materials may be used depending on what is available in the local area. Aggregate size should not be too large for your equipment. There should be a full range of sizes from dust to ½ in. or ¾ in. Some pits remove ⅜ in. and ⅝ in. material for asphalt. If there is some doubt, insist on state specification road base material.

Delivery of base material

The base material should be delivered right after the site is excavated and dumped close to the excavation or on the placed geotextile. Place and compacted it as soon as possible.

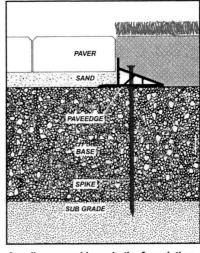

A well prepared base is the foundation for the segmental pavement.
(Figure 12-1)

Base delivery on top of geotextile fabric.
(Figure 12-2)

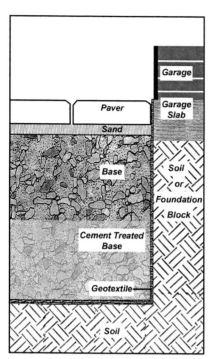

(Figure 12-3)

Base thickness

Base thickness depends on several factors: intended use, type of traffic, properties of the soil sub grade, and climate. Follow industry recommendations for all these factors to determine the proper base depth. Our example follows industry standards for a 8 in. base for driveways and 4 in. base for sidewalks.

Garage slab base adjustment

A difficult subgrade area is directly in front of the garage slab. This area is notorious for settling over a long time, warranting extra preparation. To assure adequate subgrade compaction it is necessary to excavate an extra 8 in. into the sub grade directly in front of the garage slab. Extend the excavation 3 ft., then gradually over the next 3 ft. length, bringing up to the depth of the sub grade. Compact and place aggregate base. This extra 8 in. of compacted base will help support the base in this sensitive area.

Cement treating of aggregate base:

A quick rule of thumb for cement treating base is one bag of portland cement (80 lbs.) for every ton (2000 lbs.) of base. This is a little over 3/4 cubic yard of loose crushed limestone. If treating a large area, use bobcat bucket to thoroughly mix. Add water only after spreading and raking flat.

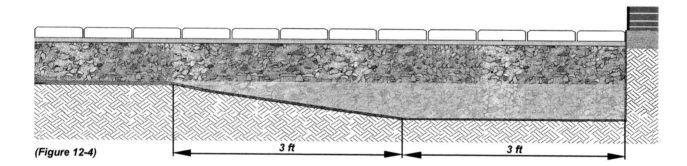

(Figure 12-4)

3 ft   3 ft

First lesson of the day.
This is a "hole".
(Figure 12-5)

Lifts

Lifts are the layers of base material that are compacted at one time. The thickness of each lift depends on the size of the compaction equipment and soil type. Do not try to compact too thick a lift for your equipment. If the lift is too thick, the compactor will only compact to a certain depth creating a 'crust' over uncompacted material. The first few inches are hard but the material below this crust is not compacted.

If using a 5 hp plate compactor, you can only compact the aggregate base in loose 3 in. to 4 in. (76 mm. to 100 mm.) lifts. This does mean multiple passes though.

Preparing for compaction

V-lines

These lines make is easier to visually locate the elevation necessary for each lift when placing and grading the base. V-lines are usually marked on a permanent structure such as a foundation or curb. Measure and mark the depth of the lifts necessary to build the base.

Pull string lines to establish the grade and thickness before compaction.

V-Lines are marks to show depth of lift
(Figure 12-6)

Adding moisture

Place the base material for the first lift. Rake it flat and smooth. Wet the base. It should not be soaking wet or have standing water, just enough moisture should be present to compact well. Dry material cannot be compacted properly. The aggregate base material should form a ball in your hand when squeezed, without crumbling or dripping water. Make sure that the moisture is completely through each layer or lift before compaction.

Keeping the base flat is an ongoing process, ideally, one crew member is raking and smoothing while another is compacting.

Check base elevations during this time. The layer should have a uniform thickness within 1". Starting with the correct base depth for the job is vital. When aggregate base material is compacted in layers, the work goes faster with the final grade easier to achieve.

Watering the base only if it is dry.
(Figure 12-7)

How do you spell "De-water"? Pump excess water out of a job rather than trying to cover with base and compacting. This would leave a soft spot.
(Figure 12-8)

Raking base.
(Figure 12-9)

COMPACTION FACTORS AND EQUIPMENT

How compactors work

Compaction equipment compacts from the bottom up, as force is applied to the soil or base the impact goes through to the harder, more resistant surface below then returns upward. The force causes the particles to move and become more dense by displacing the water and air in the material. As compaction occurs, the compactor's impact will result in continuously shorter depths until more force will return to the surface. This will cause the compactor to bounce higher and higher or to "crab" sideways making it difficult to travel in a straight line with the compactor. This usually indicates optimum compaction is being achieved.

Compactors

There are three kinds of compaction equipment: ramming, vibration, and vibratory rollers.

Ramming

Ramming machines break down and compact the soil by forcing air out from between the particles and squeezing the particles closer together. They have a low frequency, from 800 - 2,500 blows per minute, and a higher amplitude (higher stroke), from .5 inch to 3.5 inches (15 mm to 90 mm).

Jumping Jack (Hand-held upright rammer.) The Jumping Jack uses a spring mechanism (high amplitude). The jumping jack is typically used on utility trench lines that cross the excavation. It has a stronger compaction force than plate compactors. The jumping jack compactor is most effective for clay and deep areas. Other areas where the jumping jack is essential are around foundation walls, and curb lines where other equipment cannot reach.

Jumping Jack tamper (Hand held upright rammer).
(Figure 12-10)

Plate compactor (Self-propelled rammer, medium amplitude) This type has eccentric counter-rotating weights, which deliver the ramming force and travel speed.

Example: *A 5-horsepower plate compactor is limited to about 4 inches of loose base material at a time. Typically, a double drum vibratory roller compactor will handle 4 in. to 8 in. of loose material.*

Vibration (plate compactors)

Vibration machines have a higher frequency, from 2,000 to 6,000 blows-per-minute, and a lower amplitude. These are typically 3.5 horsepower to 8 horsepower. As the eccentric shaft rotates inside the vibration plate or roller shaft, it sends a stress wave into the ground. The vibration begins to move the soil particles or aggregate base material. It breaks them down, rearranges the particles, and forces the air out.

Plate compactor (self-propelled rammer).
(Figure 12-11)

Vibratory rollers (double drum, smooth drum)

Vibratory rollers work well. They are fast and very efficient. Usually recommended on jobs larger than 1,500 square feet. Units with 3,000–5,000 pounds of force will compact 4"–6" of loose base. For those with 5,000–8,000 pounds of force, 6"–8" of loose base material can be compacted. The roller action actually helps smooth the base. These are typically high frequency, low amplitude machines.

Static rollers

Static rollers apply only weight for compaction. Normally used on asphalt, these are not recommended for compacting paving soil subgrade or aggregate base material.

5 HP plate compactor.
(Figure 12-12)

Compactor Chart

COMPACTOR	BASE	SUBGRADE	
Type	Granular Base Lift Thickness	Sand	Clay
4000 lb. Forward Plate	3 – 4 in.	2 – 6 in.	1 – 2 in.
6000 lb. Forward Plate	4 – 5 in.	2 – 8 in.	2 – 3 in.
13000 lb. Reversible Plate Rammer	6 – 8 in.	6 – 12 in.	3 – 4 in.
Jumping Jack	6 – 10 in.	6 – 12 in.	4 – 6 in.
3000 lb. + Double drum walk behind roller	2 – 3 in.	2 – 4 in.	1 in.
7000 lb. + Double drum vibratory ride on roller	4 – 5 in.	4 – 6 in.	2 – 3 in.
Hand Tamper	3 – 4 in.	2 – 4 in.	2 – 4 in.

Compactor Chart
(Figure 12-13)

Large Plate rammers are a must for even residental base compaction.
(Figure 12-14)

Double drum vibratory compactor.
(Figure 12-15)

Compaction Directions

When compacting overlap each pass 50% of the plate or roller.

1. Compact the perimeter of the base, working towards the center.
2. Work from the bottom of the grade to the top in lateral passes.

   When using a vibratory roller compactor only use the vibration going uphill.
3. Compact the excavation at a 45 degree angle right (diagonal compaction).
4. Reverse the compaction to 45 degrees left.

When the lift is fully compacted, add material for the second lift. Spread evenly and compact as before until the base has reached full compaction.

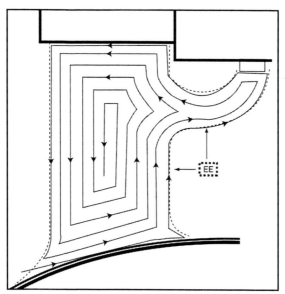

Step 1 - Perimeter compaction
(Figure 12-16)

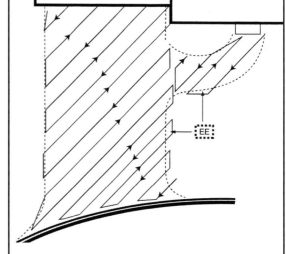

Step 2 - Lateral compaction
(Figure 12-17)

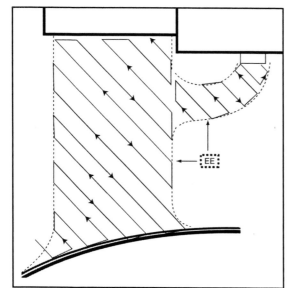

Diagonal compaction-1
(Figure 12-18)

Diagonal compaction-2
(Figure 12-19)

Over Compaction

Road builders talk of over compaction. Over compaction occurs when the base aggregates begin to separate and the fines drop, and the coarse aggregate comes to the surface. Over compaction is caused by ride on vibratory rollers improperly run by operator, poor base aggregate or a combination of both. Residential paver contractors will probably never see this situation.

As soon as the loader spreads material, start raking it out evenly. This will increase the effectiveness of the loader and compactors. (Figure 12-20)

*Rake and Compact
(Figure 12-21)*

*Rake and Compact
Always try to rake out ridges and dips in front of the compactors.
(Figure 12-22)*

Base rutting due to poor compaction
(Figure 12-23)

Always check your elevations/lifts
coming up.
(Figure 12-24)

Grading

GRADING

Grading brings the aggregate base to its final elevation.

Bobcat or Loader

Typically, a Bobcat or loader is used to make the final grade. Before the final grade double check the tire pressure, grease pins and inspect the cutting edge of the bucket. The bucket edge should be sharp to cut clean. If needed, sharpen with a grinder.

The grading process is difficult to teach because it requires a 'feel' that comes from experience. The final grade reflects the surface of the installed pavement. All transitions should be smooth, even and reflect the desired grade. Things to check before operating machine:

- good matched tires all at proper and even pressure
- greased pins for smooth operation
- sharp bucket edge (grind if needed)

Use a grinder with a metal abrasive wheel to sharpen the bucket for grading.
(Figure 13-1)

Proper base compaction

The final grade is done when the final lift of aggregate base material is placed. Bring the final lift of base material to the top 'V-line' and fully compact.

General grading and spreading is usually done by using float position on lift arms and back dragging.
(Figure 13-2)

Finish grade with a machine is usually done with a final forward cut.
(Figure 13-3)

Uniform compaction is critical for the loader to have consistent traction, if a tire hits a soft spot, the machine will buck or lurch and require another pass. Loader buckets need consistent resistance to make a cut. As you near final elevation it is easier to achieve by making a forward cut with the bucket. This requires a little extra aggregate on the final lift and but speeds up the finishing.

Setting stringline for checking grade with a base rake. (Figure 13-4)

Reference checks

Check all the elevations according to the layout of the pavement. It may be necessary to replace a lost offset with the help of the triangulation chart. Reshoot the elevations and drive grade stakes into the base. Determine if you are plus or minus on the elevation for final grade.

Hand work

Hand working the base is crucial for an even paver surface. Using a base rake and shovel, hand work the base, working no more than 1/2" at a time. Fill low spots and take off high spots to make a smooth grade. Edges, corners, approaches and curves need hand work because of the difficulty of using the loader in tight areas. Visually inspect the base. Check for irregularities by getting down on the level of the base and looking across it for high and low spots.

If you work the material too much, you will get separation with the large aggregate on the surface. If this happens, shovel off coarse material and keep working. String lines can be placed on the grade stakes 3" +/- above final grade. Use a base rake to adjust high and low spots on the base (Figure 13-4).

Low spots: Fill with aggregate base material. If it is very close, rake out the coarse rock first and spread fresh aggregate base material, rake and compact.

High spots: Scape off the extra material with a Bobcat or rake. If too much is taken off, add aggregate and re-compact.

Using Base Rake for adjusting high and low points (Figure 13-5)

PAVE TECH ©

Transitions

The key to a great looking project is making smooth transitions between elevations. Shaping the base requires an 'eye' for the look of the driveway. All transitions should be smooth and not abrupt.

The Finished Grade

The final grade sets the shape and contour of the completed pavement surface. Adjustments of contour and close attention to dips and high spots will assure a great looking pavement.

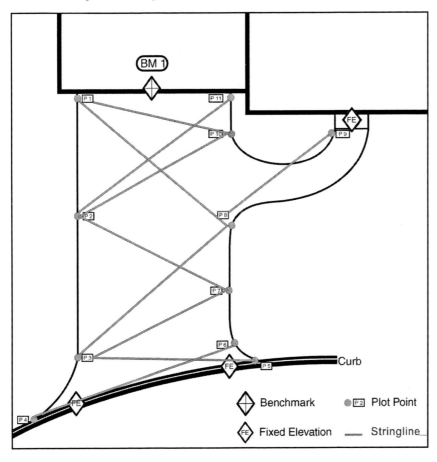

By using stringlines from Plot Point to Plot Point as shown, you can visualize the surface contour and rake it out, like in Figure 13-5.
(Figure 13-6)

During compaction of the lifts of aggregate, it is important to keep each layer flat for uniform compaction. Plate compactors will cause a defect to magnify after each pass, so it is important to follow with a base rake. Vibratory roller compactors are helpful because they actually help smooth and flatten the base, although the they can cause humps and dips at the very beginning and end of each run.
(Figure 13-7)

Compaction
(Figure 13-7)

Chapter 13 • Grading

Roll a 10 ft pipe over your surface to see if you meet 3/8" high or low tolerance.
(Figure 13-8)

Use a screed board or a rake to strike off excess material when filling low spots.
(Figure 13-9)

Low spots on the base must be filled and compacted prior to spreading bedding sand.
(Figure 17-10)

14

Edge Restraints

PERIMETER INTERLOCK

Every component of an interlocking paver system is important to ensure proper interlock, stability and durability of the pavement. Edge restraints insure interlock at the perimeter of the pavement. Edge restraints prevent pavers from excessive lateral movement. If the edge of the pavement moves, the pavement will start to fail.

The edge restraint is placed outside the pavers on top of the fully compacted base. Generally the edge restraint is installed first, reflecting the final shape and design of the project. If any adjustment to the design need to be made, it can be done at this time. An example of this may be a homeowner wanting a wider sidewalk.

When spiking the edge restraint into the base it should go into a fully compacted base extension. If the spike goes in easy and moves around, it is necessary to recompact the base. It is a lot easier to do it now prior to placing sand and laying pavers. Make all base, design or other adjustments before screeding sand or laying any pavers.

Edge restraints are critical for perimeter interlock. (Figure 14-1)

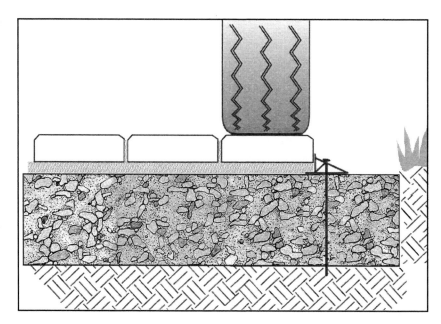

Proper installed edge restraints.
(Figure 14-3)

Quality edge restraints eliminate potential pavement deformation and failure. They are designed to stay in place, even when a tire rolls over them. They are part of the pavement system, and move with the pavement under load and with freeze thaw cycles.

- Edge restraints are positioned against the pavers to hold the paver installation in place. They are an integral part of the pavement structure. If the paver joints open up past a certain point, interlock is lost and pavement will fail.

- Edge restraints also help capture and hold the bedding sand in place beneath the pavers. Without a restraint, the sand will creep into the adjacent soil, allowing pavers to settle unevenly.

- An advantage to installing the edge restraint first is it can be used as a screed and marking guide. Using the edge restraint eliminates the need to place a separate screed guide, remove it after screeding, and fill in the void with sand after removal.

Edge Restraint Options

To maintain interlock and prevent pavement failure, edge restraints must be used on all edges of a segmental pavement. Edge restraints can be a permanent structure such as a building, garage slab, curb or a properly installed PVC edge restraint. Edge restraints are placed on top of the compacted base. They may be visible or hidden.

Edge restraints include a variety of materials. Selecting the right application for the job is important. Applications include PVC or plastic edging, concrete curbs, precast and cut stone, poured in place concrete curbs, metal edgings, troweled concrete, and treated wood. When to use the different types of edge restraints depends on the layout of the project.

PAVE EDGE edge restraint does not move or shift under even heaviest load.
(Figure 14-4)

PVC and Other Plastic Edge Restraints

PVC and other plastic edge restraints install quickly, will not rust or rot and allow for design considerations. PVC or plastic edging comes in different weights for different applications. PVC edging such as PAVE EDGE Flex is made for curves, PAVE EDGE Rigid is designed for straight areas and PAVE EDGE Industrial is designed for heavy commercial use. It is important that the type of PVC or plastic edging is made specifically for pavers. Edging made for gardens and planting beds is not the same and will not work. Use only edge restraints designed for pavers that have a reinforced design.

1. Select plastic edging appropriate for the pavement use and design. The entire edging must lay directly on the base. A key element for paver edging is the lip. The lip is placed under the paver and sand. In addition to holding the pavers in place and being integral to the pavement system, the lip also prevents the bedding sand from migrating.

Curves and straight edges are possible with PAVE EDGE.
(Figure 14-5)

2. Steel spikes anchor the edging. Use spikes that are 10 in. (250 mm) long and 3/8 in. (10 mm) in diameter. They can be either smooth or spiral. Using longer spikes does not add strength to the edging. For PAVE EDGE Rigid it is recommended that spikes be placed every 2 ft. for driveways, every 3 ft. for patios/sidewalks and every 12 in. for commercial pavements. For PAVE EDGE Flex one spike is recommended for every back support section, approximately every 8 in.

Typical spike for installing edge restraint 3/8" diameter x 10" long.
(Figure 14-6)

3. When storing PVC or other plastic edge restraints, do not store in direct sunlight. When the edging is installed properly, it will be covered, hidden from view and potential UV deterioration. PVC is recommended over other plastics because it is the strongest structural plastic and holds its shape.

4. For commercial and industrial purposes use industrial designed edging products.

5. PAVE EDGE is normally installed before screeding the sand and laying the pavers. This method is faster and more efficient than installing the edge restraint after pavers have been laid. Installing the edge restraint first allows for the exact placement of the edging according to the layout and also reduces clean up and lawn repair. When screeding, use the PAVE EDGE as a screed guide on one side and a screed rail on the other. Placing the edge restraint on the base first also speeds up marking and cutting by allowing the use of special marking tools such as a *PAVERSCRIBE* or *QUICKDRAW*.

Using the edge restraint for a guide. QuickDRAW is a great tool for marking pavers to cut for a soldier course.
(Figure 14-7)

Although edge restraints are recommended to be installed before pavers are placed, certain PVC and plastic edging can be installed after installing the pavers. This is generally a highly inefficient method. To install the edging after laying the pavers, mark, cut and replace each paver. This method results in extra work to build a larger base, screed extra sand and lay the extra pavers.

PVC edge restraint.
(Figure 14-8)

Installing PAVE EDGE after pavers have been laid.
(Figure 14-9)

If necessary to install or adjust the edge restraint after laying the pavers, first pull away the excess bedding sand. Using a trowel or shovel, cut straight down the side of the pavers to the base and pull back the sand (see Figure 14-12 below).

The bedding sand if it is to dry can be lightly sprayed with water first to help the sand stay in place. Do not scrape up the base material.

Second, place the edge restraint flat on the base and push the lip under both the sand and pavers (see Figure 14-13 below). Use a hammer to tap the back edge of the edging until it is tight to the pavers. Fill any small gaps between the edging and the pavers with sand.

Geotextile fabric can be placed over the joints of the edge restraints to prevent sand loss When laying the geotextile, overlap the joints in the edging by a minimum of 12 in. (150 mm).

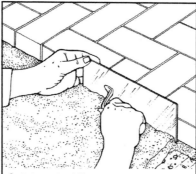

Cut down bedding sand to the base with a trowel.
(Figure 14-12)

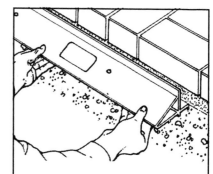

Slide the lip under the bedding sand and pavers.
(Figure 14-13)

A soldier course following installed PAVE EDGE.
(Figure 14-10)

PAVE EDGE installed with natural stone.
(Figure 14-11)

Steel and Aluminum

Steel has a disadvantage when being used as an edge restraint with pavers. Some steel easily rusts, which can spread onto the pavers. Metal edging is heavy and difficult to handle.

Aluminum 'L' shaped edge restraint lacks vertical support. When bent it does not return to its original shape and is difficult to spike. It also lacks a lip to prevent sand migration under the edging. The advantage of aluminum is it bends easily for curves and shapes, and is lightweight.

To install an aluminum or steel edging, place the smooth side up against the pavers. Aluminum and steel edging are manufactured in different thicknesses. If the pavement will carry vehicular traffic, use the thickest available. Aluminum and steel edging are staked into the base course for maximum support.

Strips of metal can also be used for edging. The strip is imbedded into the base with a supporting stake. This type of edge restraint often times will have problems with heaving from freeze and thaw cycles.

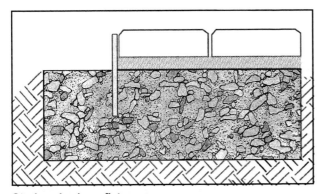

Steel or aluminum flat.
(Figure 14-14)

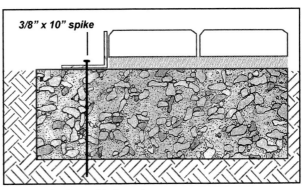

Steel or aluminum 'L' shaped.
(Figure 14-15)

Concrete Curbs, Precast, and Cut Stone

Curbs, precast and cut stone edge restraints are a visual part of the whole installation. They are all installed in a similar method with a concrete bedding or concrete backfill. If the completed pavement will receive a lot of vehicular traffic, the curb should extend to the bottom of the base.

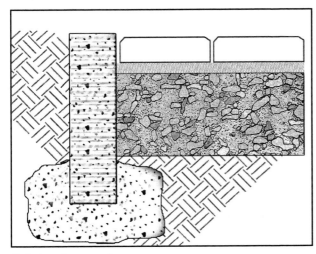

Cut stone in concrete.
(Figure 14-16)

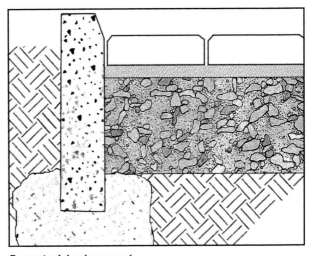

Precast edging in concrete.
(Figure 14-17)

Concrete Curbs Poured in Place

Concrete curbs or combination curbs and gutters can make a good edging for pavers. If using a poured curb to restrain the pavers, the bedding sand butts up against the back of the curb and can potentially seep through any curb expansion joints. The curb may also contain seep holes. To prevent loss of sand, place a non-woven geotextile fabric at these areas, up to 12 inches on either side of any opening. Extend poured curbs below the finished base elevation.

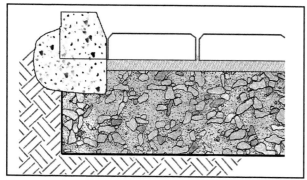

Poured in place concrete curb.
(Figure 14-18)

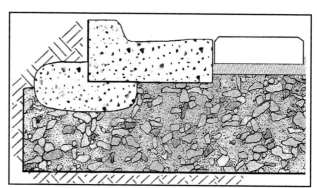

Shallow poured curb.
(Figure 14-19)

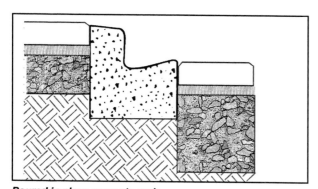

Poured in place concrete curb .
(Figure 14-20)

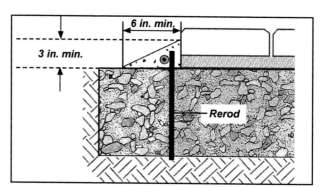

Troweled in place concrete with rerod.
(Figure 14-21)

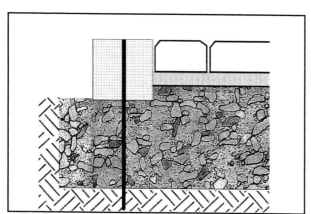

Timber edge restraint with rerod
(Figure 14-22)

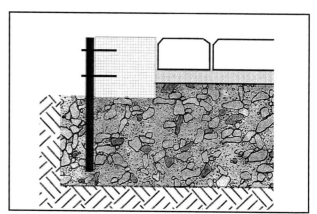

Timber edge restraint with wood side stake
(Figure 14-23)

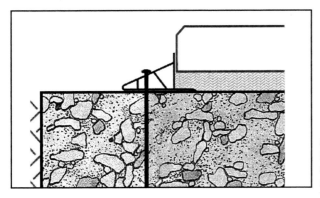

Segmental pavement
with PVC edge
restraint.
(Figure 14-24)

Troweled Concrete

Troweled edging is used when the pavers will receive light use. Regional climate is a factor when choosing troweled concrete as an edge restraint. Be aware that in areas where the temperature drops below freezing, the concrete may crack and require more maintenance than other edgings. Use concrete from either a bag mix or redimix. No forms are needed to apply it. It is simply troweled against the edge pavers and sloped downward. Insert rerod to run along the edge of the pavers for support. The concrete is placed directly on the base with staking every 2 feet. The bottom width should be a minimum of 6 inches (150 mm) wide. Minimally, the width should be the same as the depth of the paver and bedding sand.

Troweled in place concrete. Over time the pavers have pulled apart and the concrete has crumbled.
(Figure 14-25)

The edge restraint outlines the shape of the pavement.
(Figure 14-26)

Quality flexible edge restraints help with layout by providing enough resistance to create smooth curves
(Figure 14-27)

Using a screw gun and 1¼" Sheet Rock screws to permanently attach multiple pieces of flexible edging. This aids in the process of layout.
(Figure 14-28)

Treated Wood

Wood is not a permanent edge restraint. When placed in the ground where moisture and insects can get at it, wood will rot and disintegrate. If wood is used as an edge restraint make sure that only treated wood is used. Treated wood will resist insects and rot longer.

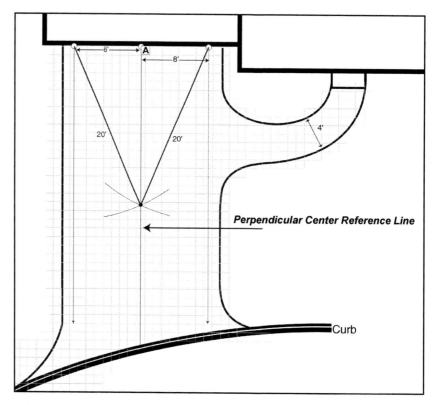

Installing Edge Restraints

In our example pavement site, four kinds of edge restraints will be used: the poured in place curb at the street, PAVE EDGE rigid for the long straight side of the drive, the garage slab, and PAVE EDGE Flex for the sidewalk and driveway entry.

The base is prepared and brought to grade as discussed in Chapter 12 and 13. The curb and garage slab are permanent structures and cannot be adjusted.

Perpendicular Center Reference Line

Straight lines are rarely found on job sites. It is often necessary to create straight lines. These reference lines are normally parallel and perpendicular to the architectural lines of a structure.

The most important reference line on the job is the *perpendicular center reference line*. Edging, bond lines and reference lines are determined from it. The perpendicular reference line is perpendicular to the structure the pavement will meet. It is often the centerline, but in the case of a turning pavement the line will be off center but still perpendicular.

To find the perpendicular center reference line, here is a field technique that is very useful. Figure 14-29 illustrates the technique.

- Determine the center of the main body of pavement. In our example, the driveway abuts the garage slab. The driveway will be 19 ft. wide. The center would be at 9-1/2 ft. from the edge.

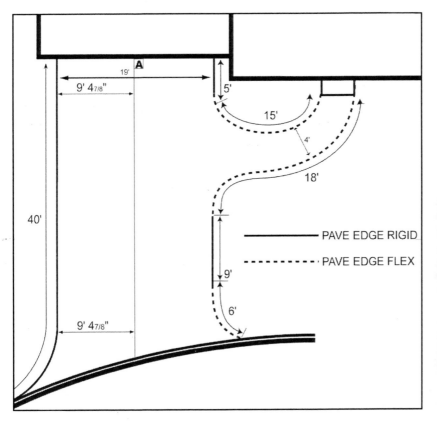

Parallel reference lines.
(Figure 14-30)

The transition from using PaveEdge Rigid to PaveEdge Flex should not be at the exact point where the curve begins. Make sure the flex edging extends into the straight area 12 to 18 in. to avoid any kinking at the joint.

PAVE EDGE Rigid can be forced to flex for gradual radius. The resistance to flexing allows for a smoother curve.
(Figure 14-31)

(Figure 14-32)

- On the garage slab measure 8 ft. from the center point to each side of it. Mark these two points.

  (Note: this measurement can be any number, but it is important to use the same measurement on each side of the center point.)

- Make two measurements of equal length out to the point where they cross each other.
- Mark where the two measurements intersect. This is the center point.
- Snap a chalk line using the center mark of the garage slab and the point where the two lines meet. Extend this line to the curb. Mark on the curb. This line is the *perpendicular reference line.*

Mark the Perpendicular reference line with either stringlines or a chalk line.

Parallel reference lines are made as needed to adjust bond lines across the pavement as needed. In our example, the edge restraint on the left is parallel to the perpendicular center reference line.

Installing edge restraints

In our example, the curb and garage slab are permanent and cannot be adjusted. The long straight edge of the pavement will be spiked in permanently at this time. Any adjustments for paver width can be made on the shorter side.

The long straight edge of the pavement is designed to start at the end of the garage. From the center reference point, measure this distance. The edging we are using has a sand retention lip. To see the chalkline when we lay the edging, it is necessary to subtract the width of the lip.

Example: 9' 6" − 1⅛" = 9' 4 ⅞"

Using this measurement, measure out from the center line down the length of the driveway. Snap a stringline to mark. The edge restraint is placed along this parallel reference line.

This photo sequence shows marking the base with paint then measuring over to compensate for the edging lip. Snap a chalkline; rigid edging would be installed along chalkline.
(Figure 14-33, 14-34, 14-35)

(Figure 14-33)

(Figure 14-34)

(Figure 14-35)

Rigid or Flexible?

The 'rule of thumb' to determine when to use a rigid or flexible edge restraint is: when the curve is more than 2 in. for every 2 ft. use a flexible edge restraint. If the curve is less than 2 in. for every 2 ft., the extra support and durability of the rigid edge restraint is better.

PAVE EDGE Flexible is good for a sharp radius. In our example site the flexible edge restraint will be used on the sidewalk and at one of the entry 'flares' of the driveway.

The transition from using PAVE EDGE Rigid to PAVE EDGE Flex should not be at the exact point where the curve begins. Make sure the Flex edging extends into the straight area 12 to 18 inches to avoid any kinking at the joint.

Rigid and flexible edge restraints are easily connected with a connecting tube. If the pavement calls for a very curvy edge, the flexible edge restraint may need to be connected with screws.

The lip

The lip is an important part of the edge restraint system. The lip is positioned firmly on the base when the edge restraint is installed. It will help keep bedding sand in place. It also allows for the edge restraint to move with the pavement during freeze/thaw conditions. Without the lip the sand will creep out from under the edges, causing pavement failure.

Spikes

Most PVC edge restraints require spiking to stay in place. The most recommended spike is a ⅜ in. x 10 in. steel spike. It keeps the edge restraint in place when hammered into the compacted base.

Spiral spikes can also be used. They are lighter and therefore cost less per 50 pound box. The disadvantage to a spiral spike is that it has a tendency to pull up base material if it needs to be relocated.

In a corrosion test, a ⅜ in. x 10 in. steel spike (not spiral) was used in a limestone base. After 7 years it was removed for testing. In this highly aggressive base material the corrosion was substantial. A wire wheel was used to remove the corrosion. Measurements concluded that it would take 60 years for the spike to lose half of its diameter under similar circumstances.

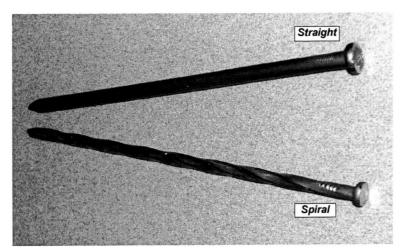

3/8" x 10" Steel Spikes
(Figure 14-36)

15

Alternate Methods of Installing Edge Restraints

INSTALLING A SEGMENTAL PAVEMENT OVER EXISTING ASPHALT OR CONCRETE

In some cases pavers can be laid over existing asphalt and concrete pavements.

If the existing pavement is in good shape with a firm base under it, pavers can make a very attractive, refurbished driveway, patio or roof deck. Check for cracks and any problems with the base prior to deciding to lay pavers over the existing pavement.

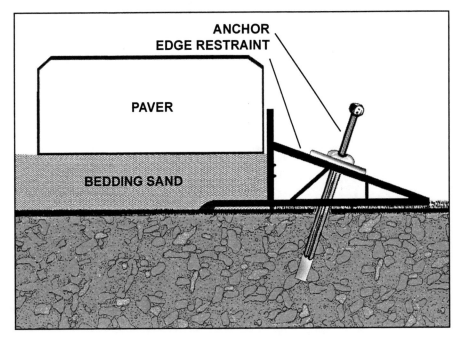

ANCHOR
EDGE RESTRAINT

PAVER

BEDDING SAND

Diagram of installation of anchor and edge restraint on existing concrete pavement. (Figure 15-1)

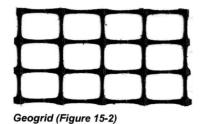

Geogrid (Figure 15-2)

The greatest concern with placing pavers over an existing pavement is loss of the bedding sand through cracks or joints. In most cases, a non-woven geotextile is placed between the existing pavement and the bedding sand to prevent this.

The other concern when constructing a segmental pavement is the proper anchoring of the edge restraint system. When using PAVE EDGE, use a NAILING lead anchor, which is a one piece nail drive anchor with a mushroom head. This will anchor the edge restraint to the existing pavement and assure proper perimeter interlock.

2 in. and 1½ in. anchors with washer
(Figure 15-3)

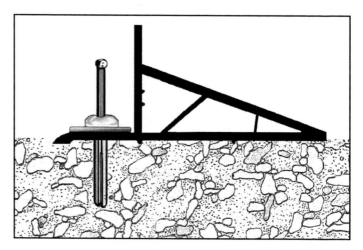

(Figure 15-4)

When installing pavers over an existing pavement it is a good idea to use a filter fabric to stop any sand from migrating into cracks and allow a path for water to exit beneath the edge restraint.

Proceed with installation, screeding one inch of bedding sand and lay the pavers.

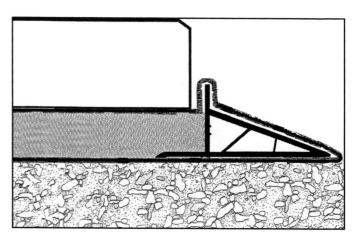

(Figure 15-5)

"Florida Method"

Installing edge restraint on sand

In certain areas of the country a proper aggregate base cannot be prepared. This is not the recommended method of paver installation.

There is an alternate method to assure a successful interlocking pavement when a proper base cannot be constructed. Without the edge restraint firmly in place the pavement will fail. This method of installation insures perimeter interlock but *is not* the preferred installation method.

The base area for the pavement is prepared and leveled. In most cases when this method is used the base and bedding sand are the same material.

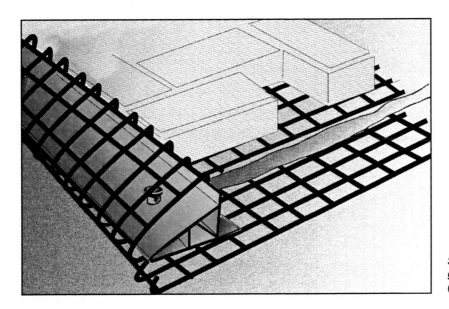

Screed 1" of bedding sand over the geogrid and lay the pavers. (Figure 15-6)

Geogrid is cut into sections 2 ft. x 6 ft. These are laid at intervals at the edge of the pavement (see Figure 15-7). They should extend at least 3 ft. - 4 ft. into the pavement area and under the bedding sand.

Place the PAVE EDGE around the perimeter *over* the geogrid.

Sand is screeded and pavers are laid as normal.

As the pavers near the pavement edge, the geogrid is wrapped over the edge restraint back into the pavement area on *top* of the bedding sand. The pavers are then laid on top of the geogrid. The pavers will anchor the wrapped geogrid.

A segmental pavement done is this manner is solid and will set up interlock during compaction.

Lay sections of geogrid material on the perimeter of the area to be paved.
(Figure 15-7)

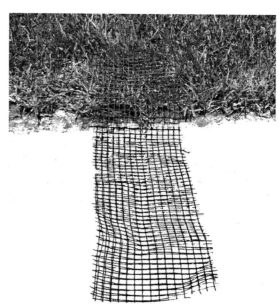

Cut the geogrid material into sections 2 ft. x 6 ft.
(Figure 15-10)

Screed 1 in. of bedding sand over the geogrid and lay the pavers.
(Figure 15-8)

Wrap the geogrid around the edge restraint and lay into the pavement area.
(Figure 15-11)

The pavers will hold the geogrid in place which hold the edge restraint in place.
(Figure 15-9)

Pavement interlock is set up with the final sweeping and vibration of the sand in the joints.
(Figure 15-12)

16
Sand Properties

EVALUATING SAND FOR BEDDING AND JOINTS

As the contractor or installer of a flexible pavement system it is your responsibility to know the materials you need to do a job correctly. One of the key elements of a flexible pavement system is the bedding and jointing sand.

The sand should contain a range of particle sizes having the gradation necessary for the flexible pavement system. More than one type of sand can be used providing they provide the full range of gradation between them.

Gradation

Measuring sand suitability for paver applications is called gradation. When purchasing sand from your supplier it is up to the contractor to know what he is buying. In different areas of the country, proper sand gradation is called different things. To assure that the sand you are using meets with the American Society for Testing and Materials (ASTM) standards, ask the aggregate supplier, pit or quarry for a gradation certificate and test report. If one is not available take a sand sample to a local materials testing lab. It is critical to use the correct sand for both the bedding layer and for the paver joints. If the sand is either too coarse or too fine, it will not do the job properly. It may take a bit of investigation with your local suppliers and gravel pits, but it is well worth it

Joint sand prior to compaction (Figure 16-1)

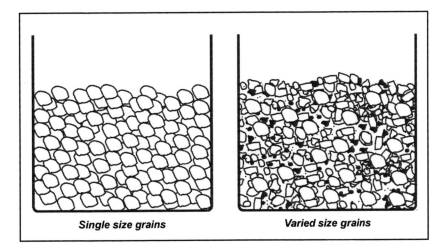

Single size grains   *Varied size grains*

Gradations of sand (Figure 16-2)

	GRADING FOR JOINT SAND – ASTM C-144	
SIEVE SIZE	NATURAL SAND PERCENT PASSING	MANUFACTURED SAND PERCENT PASSING
No. 4 (4.75 mm)	100	100
No. 8 (2.36 mm)	95 TO 100	95 TO 100
No.16 (1.18 mm)	70 TO 100	70 TO 100
No. 30 (0.600 mm)	40 TO 75	40 TO 75
No. 50 (0.300 mm)	10 TO 35	20 TO 40
No. 100 (0.150 mm)	2 TO 15	10 TO 25
No. 200 (0.075 mm)	0 TO 5	0 TO 5

ASTM C-144

ASTM C-144 is the standard for sand in masonry mortar. Sand meeting this standard is also suitable for joint sand in flexible pavement systems. After receiving the gradation certificate or test report compare the numbers to those in the ASTM chart. If the sand does not meet these standards, reject it.

Testing

Testing of sand is done by a series of sieves or screens. Each different size sieve is labeled by both a designated number, like a No. 8 and by the size of its screen openings. Look at the chart (*Figure 16-3*) to see how particle size relates to screen size.

The sieve on the top has the largest screen size, followed by progressively smaller screens. Very little will pass through the bottom one. It has a very fine screen.

When the sand sample is put in the top sieve, the screen allows only particles that are smaller than its openings to pass through to the next sieve. The same happens at the next sieve and so on until the sample reaches the bottom sieve. When the sieves are separated, a portion of the sample remains on each one.

Each portion is weighed and compared to the total weight of the sample. It is assigned a percent of the total weight and that percent is used to determine if the sample meets the American Society for Testing and Materials (ASTM) specification for bedding sand ASTM C-33, or joint sand C-144.

Sieve Size	Opening Size, Name	Percent Passing
2 inch	50.00mm, gravel	100%
1 1/2 inch	37.50 mm, gravel	100%
1/2 inch	12.50 mm, gravel	100%
3/8 inch	9.50mm, pea gravel	100%
No. 4	4.75 mm	95-100%
No. 8	2.36 mm, sand	80-100%
No. 16	1.18 mm	50-85%
No. 30	0.60 mm	25-60%
No.50	0.30 mm	10-30%
No. 100	0.15 mm	2-10%
No. 200	0.075 mm finest/dust	

Sieve size and percent passing (Figure 16-4)

(Figure 16-5)

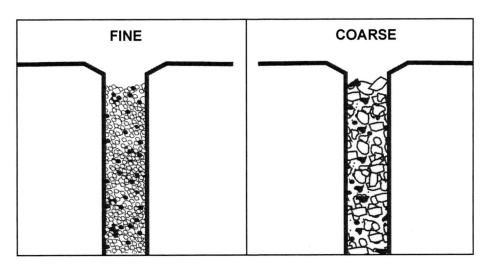

FINE	COARSE

*Joint sand
(Figure 16-6)*

Sand for the bedding layer

If possible and available, it is convenient to use the same sand for both bedding and jointing. Use coarse sand for bedding. Sand that meets this criteria might be called concrete, plaster, sharp, torpedo or a number of other names. All sand used in the installation should be sharp, clean and free of dirt, clay, dust or foreign material.

Mason sand

Do **not** use mason's sand for bedding! Most mason sands are too fine. Quite often, an uninformed installer will make the wrong choice and use mason's sand. Mason sand is a fine sand used to make mortar (sand plus cement). When placed under the pavers, mason's sand will settle at a different rate, causing an uneven, distorted appearance on the pavement surface. The small particle size will cause it to shift under load and not drain water properly.

Sand for the paver joints

To fill the paver joints, use bedding sand or a sand that meets ASTM C-144.

Using bedding sand for joint sand

Make sure the joints are packed full after vibrating. The larger particle sizes of coarse sand can become wedged at the top of the joint leaving an unfilled space between the pavers. When settling occurs, it may result in a costly callback by the customer who wants more sand swept into the joints.

A finer sand vibrates into the joints more readily than a coarser sand, but if a finer sand fills joints more quickly with less passes of the compactor and broom, it also means it will come out of the joints just as easily. Fine sand does save time when filling the joints and is less likely to settle, but more likely to come out of the joints.

Sand hardness

Under heavy traffic, soft sands may disintegrate, allowing the pavers to shift or settle. The hardest sands are naturally occurring. Nature has already worn them down to a weathered state. Sand can also be made at a quarry. Existing rock is crushed and screened to match ASTM C-33 and C-144. Critical to the success of a good sand is its hardness. Particles should not easily break or crumble.

Test hardness

A simple test to determine hardness is to take the flat of a knife blade and try to crush them on a hard surface. How easy the sand will crush indicates the sands' hardness and suitability.

> Note: Limestone sand, depending on where it is from, may or may not meet the standards. Be sure to check its analysis carefully, and field test hardness.

Avoid stone dust

Most types of stone dust are inappropriate for bedding or jointing. It usually packs too tightly, not allowing any drainage. One of the purposes of bedding and jointing material is to allow water to drain through the pavement bedding layer. Stone dust is also very difficult to use for jointing because of the elongated particle size. In many cases, stone dust is too soft a material to use. Remember stone dust is a by-product and considered by quarries to be waste.

Controlling sand moisture at the job site

Controlling the moisture in bedding and jointing sand is important on the job site. Bedding sand should contain some moisture. Bedding sand should contain just enough moisture for the sand to screed off smoothly and to hold its place while placing pavers.

Jointing sand should be as dry as possible. The sand needs to be sweep into the joints and then compacted, vibrating more sand between the pavers. Dry sand does this more efficiently.

Here are some tips for controlling sand moisture for each.

Bedding sand moisture

For bedding sand, it is good for the sand to contain some amount of moisture.

- Leave bedding sand in small piles on the job site.
- If the sand is free draining (no dust), do not tarp the bedding sand.
- If the bedding sand is rained on, just take what is needed for the job from the top of the pile not the bottom until it has had time to drain.

Placing piles of bedding sand on base (Figure 16-7)

Jointing sand moisture

Jointing sand should be as dry as possible.

Tarp jointing sand. Avoid piling sand in areas subject to rain water channeling, such as against street curbs. If necessary to pile sand in such areas, lay a piece of plastic pipe down first so water can pass.

At the beginning of each day spread approximately the amount needed for the day's work on the open pavement to dry thoroughly.

17

Sand Screeding

SCREEDING

Screeding is the action of striking off the bedding sand to an even level prior to placement of pavers. Screeding bedding sand to a uniform thickness is important for proper paver placement and elevation. During compaction the bedding sand is driven up between the joints creating the beginning of vertical interlock. The bedding sand also allows for all pavers to move to a uniform elevation during the compaction process even if the different pavers have minimal height deviations.

Tools for Screeding

Tools for screeding include: Wheelbarrows, shovels, screed rails, screed boards, trowels, hand pulls and mechanical screeds.

Screed Boards

Screed boards can be wood or metal. The screed board is pulled along the top of screed rails to strike off the sand at a uniform thickness. Wood screed boards are 2 in. x 6 in. or 2 in. x 8 in. and no longer than 10 ft. Inspect wood screed boards frequently for level. Aluminum or magnesium screed boards should be either 1 in. x 4 in. to 1 in. x 6 in. tubing.

Screed boards can be used with edge restraints. Notch each end of the board ¾ in. This will allow the board to slide smoothly along the edge restraint, screeding the sand to an even 1 in.

Placement of bedding sand before screeding. Keep sand in piles for better drainage if it rains and easier spreading.
(Figure 17-1)

Trowels

Trowels are for smoothing sand in small, tight areas. Using trowels to smooth bedding sand in larger areas to an even level is a very slow process and not recommended. When using trowels do not press down and compact the sand. Keep the trowel at an angle.

Using trowels on bedding sand.
DO NOT FLAT TROWEL.
(Figure 17-2)

Screed rails (bars)

Screed rails are used as sand height guides on the base. Sand is spread between them with the excess sand struck off with a screed. To screed sand to 1 in. in depth, a screed rail that is 1 in. in diameter is used.

Some contractors prefer square screed rails because they stack and are easier to transport. We recommend round screed rails such as conduit or water pipe, because they lay truer to the base. A round screed rail has only one point of contact on the base giving it less problems with material that may gather underneath it. The screed board strikes off the excess sand leaving a layer of 1 in. (2.5 cm) of loose bedding sand.

Lengths of the screed rails vary depending on the amount of contour in the base. On most jobs, 2 sizes, 5 ft. (1.5 m) and 10 ft. (3.0 m) screed rails are enough.

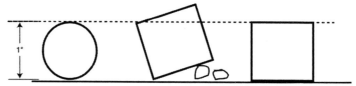

Round screed rails work better than
square but square screed rails handle
better.
(Figure 17-3)

Comealongs

Filling screed rail marks
with a SandPULL.
(Figure 17-4)

Also known as SandPULL's, these tools are long handled tools that allow workers to move excess sand by pulling. The better designed tools have one side to free float tight areas and fill screed rail marks.

GUIDELINES FOR SCREEDING BEDDING SAND

Steps in screeding

Prior to screeding the sand, check base course elevations with a straight edge or screed pipe. Make any adjustments to the base prior to screeding the sand. Do not fill low spots with more sand. Check for proper elevation and flatness.

Screeding normally starts at the bottom of the grade. If conditions allow, the bedding sand should be delivered to the top of the grade, where the last of the sand will be screeded .

Low spots on the base must be filled in prior to spreading bedding sand.
(Figure 17-5)

Laying out the screed rails.

The distance between rails is determined by the length of the available screed board and the grade and/or contour of the base.

Lay rails in the direction of the water runoff (down hill on a grade) to avoid low spots.

For flat areas, place the screed rails on the base course, parallel to each other. Place the screed rails 8 ft. to 10 ft. (2.5 m to 3 m) apart depending on the length of your screed board. Place each screed rail close enough to allow the screed board to go past each rail at least 6 in. (15 cm).

If the base is highly contoured, place the screed rails every 2 ft. to 3 ft. (60 cm to 90 cm) apart. The more contour in the base, the closer the rails need to be set. When screeding around curves or circles, place the rails in the pattern of wagon wheel spokes, screeding away from the edge. Always make sure the screed board overlaps the guides by at least 6 in. (15 cm) on each side at the widest distance between the rails. In a heavily contoured base area, you might only screed a 4 ft. x 4 ft. (1 m x 1 m) area at a time. Smaller areas are usually done by free float SandPULL's or by trowels.

Placing piles of bedding sand on base
(Figure 17-6)

Avoid changing screed rail directions abruptly. The risk is a low spot at that intersection. Keep the screeded sand even and smooth.

Placing sand

Carefully shovel or dump sand over and around the screed rails. Make sure the rails stay in place and no sand gets between the rail and the base. You should not need to recheck the elevation if the base has been properly checked.

Depending on the size and type of installation, use a bobcat, or wheelbarrow, to move the sand onto the base. Screed only the amount of sand that will be covered with pavers that day. When distributing sand on the base, place the sand in uniform piles on the base. (Figure 17-6). Leave it that way until you are ready to screed.

Use shovels or comealongs to roughly spread the sand in the area to be screeded.

Sliding down screed rails for the next screed pull.
(Figure 17-7)

Sand prior to screeding placed between screed rails (Figure 17-8)

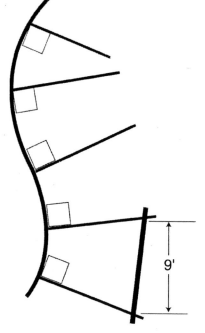

Placement of Screed Rails on curves (Figure 17-9)

Screeding

Starting at the curb, screed the 'flare' or extended areas of the driveway first.

Screed the driveway by dividing the width of the drive into screedable widths. Screed each width the length of each rail. Use proper length rails that will not raise up off the base. Screed rails must lay flat on the base. Once the screed rails are in place and sand is spread, push or pull the screed board along the top of the rails, striking off the sand as you go. One worker pulls the screed board while another uses a shovel to add or remove sand as needed.

When pulling screed rails to the next position, do not lift them. Slide them through the sand leaving 6 in. to 12 in. of the rail in the previous screeded area. (Figure 17-7).

Screed sidewalk and feather into the driveway. The sidewalk is 4 ft. wide and has edge restraints on both sides. The edge restraints can be used as screed rails with a notched screed board.

SandPULL Pro takes the kneeling out of screeding (Figure 17-10)

Two people make screeding a more efficient job
(Figure 17-11)

Screed board notched for edging.
(Figure 17-12)

Positioning screed rails in sand already spread. Wiggle and push down into the sand until it makes full contact with the base.
(Figure 17-13)

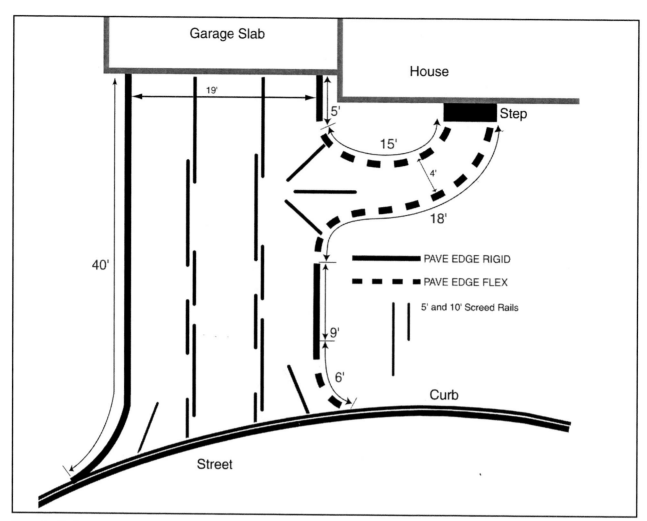

Screed rail placement and grade of
driveway
(Figure 17-14)

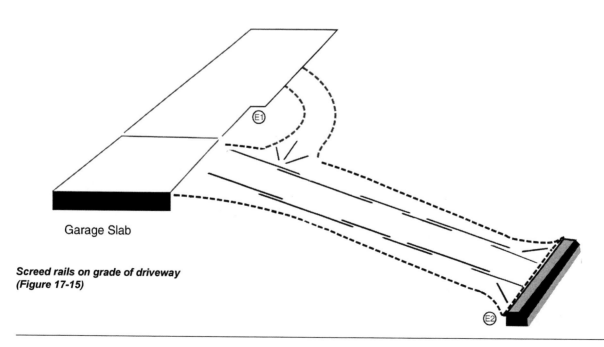

Screed rails on grade of driveway
(Figure 17-15)

Feathering

When intersecting grades such as a sidewalk and driveway, it is usually necessary to feather the bedding sand together. Smooth the transition with a SandPULL or a larger trowel about 18 in. in length.

Placing screed rails after sand is spread

Changing position or placing screed rails after sand is spread is possible. To place a screed rail in sand that is already in place; wiggle the rail back and forth with a downward pressure, placing it securely on the base with out excess material under the rail.

End of day

Some companies screed the entire amount of area that they expect to cover with pavers in one day at once. Others will screed twice, once in the morning and once in the afternoon. Leave at least 2 ft. of screeded sand uncovered by pavers at the end of the day. If all the delivered sand is not used in one day, the bedding sand can remain uncovered. The jointing sand needs to be tarped if rain is expected. If rain is expected, cover any screeded areas with plastic. Always shingle the plastic if more than one sheet is used and cover at least 2 ft. of the pavers at the laying edge. Screeding more than what can be paved in one day is not a good idea. Sand can dry out, get soaked by rain or be disturbed by passersby or pets.

Mechanical screeds for small projects

Small mechanical screeds are pulled by one or two men. They work well for small to medium jobs. Standing mechanical screeds like PAVE TECH's SandPULL Pro eliminate screeding bedding sand on the knees. Screeds like these are much easier on the back and knees of workers along with speeding up screeding.

Mechanical screeds for large projects

On large projects like streets or parking lots, large mechanical screeds can be used. A mechanical screed runs along a screed rail or curb and sand is added as needed. These can be pulled by a bobcat or loader. A larger area can be screeded at one time than with hand screeds. When using mechanical screeds, check elevations often, using the height of the screed rail to check for correct sand depth. They work best for large flat areas.

Screeding sand with SandMAX I (Figure 17-16)

*Mechanical Screed SandMAX II
Made for screeding large areas
efficiently
(Figure 17-17)*

Fixed elevation adjustments

When screeding the sand next to a garage slab, start with pavers flush with concrete or a little lower. After tamping, pavers will settle ¼ in. +/- ⅛ in. The lowered level in this area prevents snow and water backup in these areas. When coming down a grade, the paver should start about ½ to ⅜ in. high at the concrete. After compaction you should be about ¼ in. higher at the curb. This helps surface water run off. If the pavers are too low they will collect dirt and sediment at that junction.

Conclusion

After enough sand is screeded, it is time to start laying the pavers. Prevent people and animals from walking on the bedding sand, always fix disturbed sand before laying pavers on it.

18
Laying Pavers

WHERE TO BEGIN

Whenever possible paver projects start at the low end of the grade and build uphill. This is important to prevent shifting of the pavers during construction. Pavers 'float' on the bedding sand until they are compacted. Try to avoid laying pavers down a grade, it is very difficult to keep them tight. In our example, the grade goes up from the street. If the grade of the pavement went up from a building, the pavers would be laid starting at the building. You will find that in most cases the grade will drop away from a building.

Building the field

The design of our example pavement calls for a double offset soldier course at the curb of the driveway and a single soldier course to outline the entire pavement. The paver is rectangular, laid in a herringbone pattern.

A permanent edge restraint has been installed on the left side of the driveway. The edge restraint on the other side is temporarily installed. It will be staked in permanently after the pavers have been laid across the pavement determining the exact width of the pavement, allowing full unit pavers. This method eliminates excessive cutting, creating a clean edge.

Laying pavers on a driveway.
(Figure 18-1)

Laying out bond lines

Parallel Reference Bond Line

The sand has been screeded as discussed in Chapter 17.

The *parallel reference bond line* is parallel to the main structure the pavement will meet. In our example, pavers line up evenly to the garage slab. The placement of the first parallel reference bond line is usually 4 to 6 feet from the beginning of the pavement. In our example it is placed above the 'flares' of the driveway. The line of the garage slab is transferred to create the parallel. Transfer this line by measuring out 2 points of equal distance. Snap a chalk line in the bedding sand.

Parallel reference lines are created as need based on the complexity of the job layout. For our example site, we will create one at the beginning and one at the sidewalk to help keep bondlines straight.

Perpendicular reference lines

The center perpendicular reference line was used to determine the location of the edge restraints. We can use the edge restraint to determine the edge of the soldier course. The pavement design calls for a soldier course to border the entire pavement. The paver pattern will be enclosed by this border.

Define the width of the soldier course by placing 1, 2 or 3 pavers along the edge restraint on the left side of the pavement. Snap a chalk line. This line is perpendicular to the garage slab and is offset from the center reference line.

The point where these 2 lines meet is where the paver pattern starts.

Center bond line

To find the center bond line lay paver across from soldier course perpendicular reference line. Once you have established a pattern, measure the nearest bond line to your perpendicular reference line. Now take this measurement and snap a perpendicular center bond line.

(Figure 18-2)

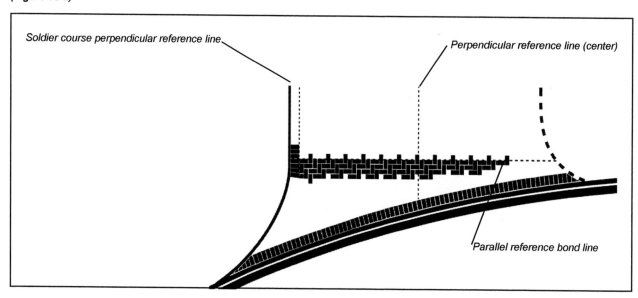

Soldier course perpendicular reference line

Perpendicular reference line (center)

Parallel reference bond line

Layout a soldier course and straighten.
(Figure 18-3)

The setter places the pavers on the parallel reference line.
(Figure 18-4)

Build back against the soldier course and curb to create foundation.
(Figure 18-5)

The first paver laid

The first pavers are laid along the curb. Curbs are rarely smooth and even. A soldier course laid along the curb creates an even, smooth looking edge. Lay one row of pavers forming the soldier course along the curb. The second row of pavers will be set later.

Start the paver pattern at the point where the parallel reference line and the soldier course perpendicular reference line meet. Lay the pavers across the pavement along the parallel reference line. Build a complete pavement between the parallel reference line and the curb. This field should be at least 3 to 4 feet thick.

The field pavers will need to be cut to lay in the second row of the soldier course. Using a *QUICKDRAW* tool, the field pavers can be marked for cutting. Remove the pavers for cutting and lay in the second row of the soldier course. Cut the field pavers and replace in the pattern. Once complete, tamp and sweep sand into the joints of the first few feet of pavers. This will act as the base or foundation for the rest of the pavement.

NOTE: *When there are large gaps because of poor concrete, use a little base material to fill and maintain the spacing. If sand is used it will generally wash out of gaps wider than 1/4 inch.*

Chalk lines or string lines

Chalk lines are recommended because they cannot be disturbed by careless feet. When there is a lot of contour to the pavement, snap a chalkline only 6 feet to 8 feet long. Overlay half the distance of the first snap and snap another. Continue this way until finished. This method should ensure a long, straight line.

To protect the location of snapped chalk lines at the end of the day or in case of rain, lay a paver out from the laying edge on top of the chalk line. To redo the chalk line, remove a paver from the laying edge that is over the original chalk line, uncovering the line under the paver. Match that with the protected line under the paver set out from the laying edge. Re-snap the chalkline.

String lines can also be used. When a stringline is used for parallel reference lines, stakes are driven into the soil on the side of the excavation and line is tied between them. If a stringline is used for the perpendicular bond line, one end of the string is tied to a stake that is staked in the base and the other is wrapped and anchored around pavers on top of already installed pavement.

Manufacturer's paver shapes and patterns

The pavement is usually designed by the contractor with the customer. *Paver shape* is the actual shape of the individual paver. The pavement design is made from paver shapes placed in a pattern.

Paver bond pattern is the way the individual paver shapes are laid in reference to one another. For complicated shapes and patterns refer to the manufacturer's information for placement.

See Chapter 4 – Paver Shapes and Patterns for more information.

Place a paver out from the laying edge over the chalk line at the end of the day.
(Figure 18-6)

Stringline can be used but are not recommended because they can easily be disturbed.
(Figure 18-7)

(Figure 18-8)

Laying pavers

A paver is laid by holding it between the fingers and thumb. The first two pavers are set to start the pattern. The next paver is brought into the corner made by two pavers. With the fingers on the outside and the thumb on the inside, it is set in place. *Do not drop the paver!*

Consistency is the word to remember when hand setting pavers. Consistent joint widths help spread loads across the pavement and also make the pavement look clean, orderly, and professional. As a beginner, do not be frustrated if your first few rows of pavers are not perfect. After a little practice, your bond lines will begin to stay straight. The soft but sure method will help you maintain consistency as you lay the pavers. Do not throw or beat the pavers into place. *No hammers.*

*Setting pavers
(Figure 18-9)*

- Hold the paver firmly and set the bottom 1-2 inches of it against the upper side of the paver already in place. Be careful not to knock the already set pavers too hard or it will shift the bond lines. Move the thumb to the top of the paver.

- Lower the paver downward while applying downward pressure with your thumb, while also applying slight sideways pressure with your four fingers. This even pressure gives good results – a consistent joint of 1/16 inch to 1/8 inch (2 mm to 3 mm). Do not try to visually space the pavers. Always lay hand tight. A strong force is not appropriate.

The person laying the pavers is called the setter. On most residential jobs, there is only one setter. It is faster to have one setter with the rest of the crew supplying the pavers to the laying edge than it is to have 2 or 3 setters always stopping and going back for pavers.

The crew leader or foreman is responsible to keep the construction moving, so should not be the setter. That person makes sure that the right blend of pavers are brought to the laying edge, that sand is screeded ahead of the setter and the entire project is moving smoothly.

*Moving pavers on a site with pallet wagon.
(Figure 18-10)*

Paver cubes, bands, stacks

In Chapter 9 Logistics, the pavers were delivered and placed along the curb in front of the property. A cube of pavers consists of 6 to 8 bands of pavers. Pavers in different cubes may have different hues and shades of the same color. To assure a mix of colors on a project, pavers are taken alternately from different cubes. Keep the job moving smoothly by taking bands of pavers from cubes as numbered on Figure 18-13. To start, take a band from cubes 1 and 2 and then go to the other side and take a band from cube 3. This not only assures an even mix of pavers, but as the project progresses, the distance from the laying edge to the pavers does not increase.

A band of pavers is brought to the laying edge with a paver cart. The bands are set back from the laying edge. Mix the pavers from each band in stacks of 8 to 10 pavers. This mixture of pavers is then carried to the laying edge. These stacks should not be too high or they will fall over. The pavers are placed even to the laying edge not more than 12 inches from the edge. Room should be left between the stacks of pavers for the setter to move easily around and straddle them. The setter will move from one stack to another laying the pavers. The crew must keep a supply of pavers on hand for the setter to keep the construction of the pavement moving. There is no set amount of pavers that one setter can lay. As the crew works together, a rhythm will develop that is comfortable for everyone.

The PAVERCART easily takes pavers off a pallet or brick stacks.
(Figure 18-11)

A band of pavers is brought to the pavement.
(Figure 18- 12)

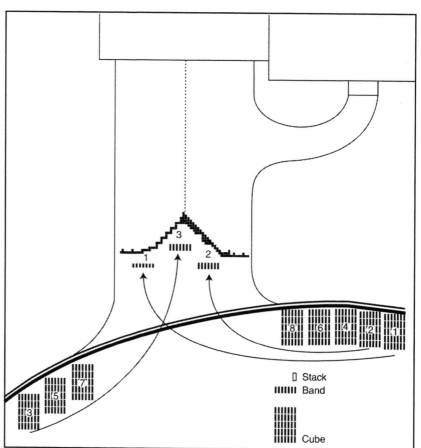

(Figure 18-13)

PAVE TECH ©

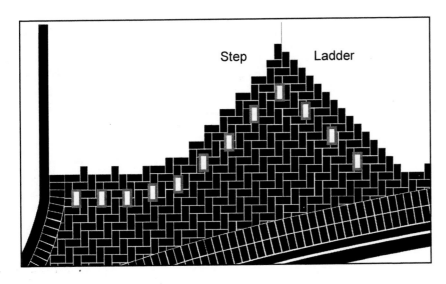

(Figure 18-14)

Laying pavers in a pyramid

Once the base field of pavers are laid, the pavement is built by working on both sides of the perpendicular reference line. The original center reference line may or may not line up to the bond lines of the pavement. Re-snap an offset perpendicular reference line. This line will be within 2 inches of the original line.

The pavement is built up evenly on both sides of this line in the shape of a pyramid. The long side of the paver is the 'step' and the short side is the 'ladder'. When setting the pavers keep an even mass or amount of pavers on each side. This balance is the key to keeping bond lines straight. If only one side has pavers laid on it, the bond lines will slowly creep to the other side of center, pushing the bond lines off. Work back and forth from one side then the other. Complete the pavers to the edges.

This method of laying also allows workers to stack pavers for laying on one side, while the setter lays the pavers on the other side.

Paver setter with crew members bringing pavers.
(Figure 18-15)

Bands brought to laying edge by Paver Cart. Then stacks set near laying edge for paver setter.
(Figure 18-16)

Adjusting bond lines (should always be done with a string line.)
(Figure 18-17)

(Figure 18-19)

(Figure 18-18)

Keeping bond lines straight

Check bond lines every 4 to 10 feet during the course of the job. Most installers first reaction to bond lines moving is to get out a hammer and start beating the pavers into alignment. That is not the best solution.

After enough pavers have been laid to fill the entire width of the pavement, check the bond lines both directions.

Set a string line to align with the original parallel bond line. Stretch it across the top of the pavers by tying it around 2 stacked pavers. If the pavement has a lot of contour, place an object under the line to lift it.

With an alignment bar, adjust the pavement by opening the joints. Come forward every 4 feet. Once the bond line is straight go back and even out the gaps in the joints.

To adjust the perpendicular bond lines follow the same procedure with a string line. This time going up and down the pavement.

Always adjust bond lines before compacting, as well as before marking or cutting.

PaverPaws and SlabGrabbers

Special tools have been designed to assist when using larger pavers. If the paver is too large for the setter to handle, tools such as a PaverPaw or SlabGrabber offer a solution. These types of tools securely grab an individual paver for placement.

PaverPaws (left) help when paver units are too large to grasp with one hand.
(Figure 18-20)

SlabGrabbers help lift and set oversized units.
(Figure 18-21)

Gloves and Finger tape

Pavers are laid by hand. Under dry, normal circumstances, hands develop callouses. In wet, damp weather, skin and callouses can peel causing problems for setters. Production rates go down if workers have problems with their hands. Some pavers also have sharp edges, that make them difficult to handle.

Taping the fingers is the easiest way to protect them. The best type of tape to use is medical tape (Finger Tape) that sticks to itself, but not to the skin. It stays on in wet weather and is easily removed. Avoid tapes that stick to the skin such as duct tape.

Gloves are also worn for protection, but unless they are tight fitting they do not work very well. The glove can get stuck between the pavers slowing down laying.

(Figure 18-22)

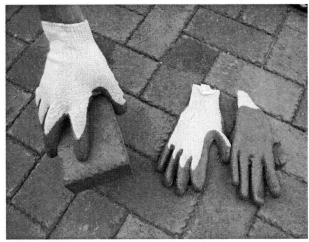

(Figure 18-23)

PAVE TECH ©

Paving in Difficult Areas

19

PAVER CHALLENGES

Placing pavers around trees, manhole covers, and other openings are challenges that are not as hard as they appear.

Paving by an opening

When paving around an object it is important to maintain perfect bond lines. The bond lines in front of the obstruction should be identical to the bond lines on the back of the obstruction.

Run perpendicular and parallel chalk lines around the obstruction. If the object is a tree, snap chalk lines parallel to each other on both sides of the tree. Make sure they are perpendicular to your base line. Then transfer the base line to the front and back of the opening. This same method is used on larger openings also. The key is to maintain the bondlines on the front, back and sides of the obstruction.

Continue to lay pavers evenly on both sides. Lay the pavers so that both sides are built up evenly so that when the back side of the opening is reached both sides are even.

When the back side of the opening is reached, fill in the pavers between. Use chalklines or stringlines to follow bond lines. Continue on with paver placement. Correct alignment with an adjusting tool if necessary.

After the bond lines are checked and corrected, cut pavers to fill any gaps or spaces that are left around the opening.

(Figure 19-1)

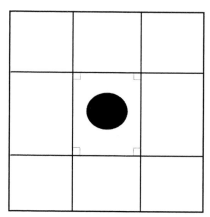

(Figure 19-2)

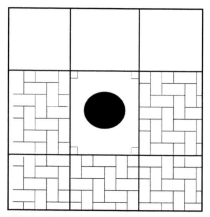

(Figure 19-3)

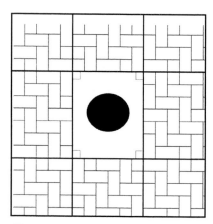

(Figure 19-4)

Paving around an opening

There are no tricks to paving by an opening, the idea is to continue to lay the pavers as if there was an imaginary edge at the opening. When paving by the opening do not increase the pavement any further out and do not decrease the size of the pavement. Complete the entire field of pavers in front of the opening following the bond pattern. Maintain the bond lines past the opening. Once you are past the opening, fill in the pavers between the two sides.

If necessary fill in the opening with any cut pavers after the pavement is completely past the opening.

For example: A manhole where it is necessary to fill in the rounded corners.

Changing direction (Figures 19-5, 19-6 & 19-7)

In our example pavement, there is a sidewalk that comes off the main body of pavers. The pavement pattern will not change because of this change of direction.

In the opening for the sidewalk, snap a perpendicular line to the main pavement center line.

The field of pavers of the main driveway act as the base for the pavers of the sidewalk, just as the bottom of the driveway was a base for the pavers on the main driveway.

The pavers are laid in the same pattern following the new reference lines into the sidewalk areas. At a curve or any change in direction, the same theory applies. Build past the opening without increasing or decreasing the pavement along an imaginary line running by the opening. Once the pavers are in place forming a base, then the turn is started by snapping a new perpendicular reference line.

Paving by non-aligned edges

Sometimes the direction you want to pave in may not be perpendicular to the starting edge.

Pavements almost always start away from and build towards a building. Normally, bond lines are run off the base line of the building. This ensures that any cutting near the building will be uniform. If your starting area is just too far, find another, closer major feature.

- Establish a straight starting line using a 90 degree angle.
- Lay a few rows of pavers.
- Check the alignment, straighten as necessary.
- To fill any gaps, cut pavers to fit the gaps.

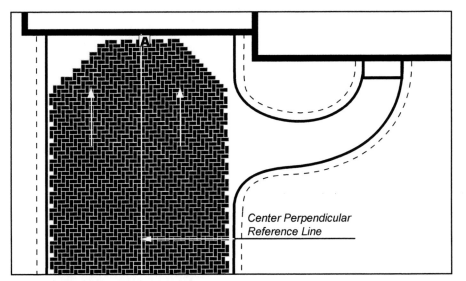

Lay body of pavers past the opening for the sidewalk. (Figure 19-5)

Center Perpendicular
Reference Line

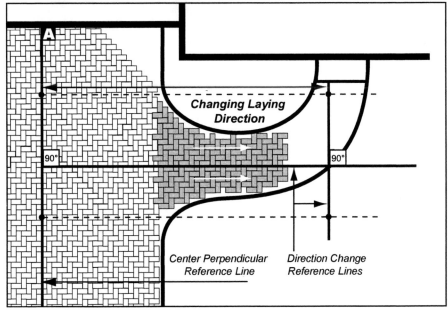

Next, create a new perpendicular reference line. This line should be 90˚ to the center perpendicular reference line and follow a bond line on the main body pavers. (Figure 19-6)

Changing Laying
Direction

90°   90°

Center Perpendicular
Reference Line

Direction Change
Reference Lines

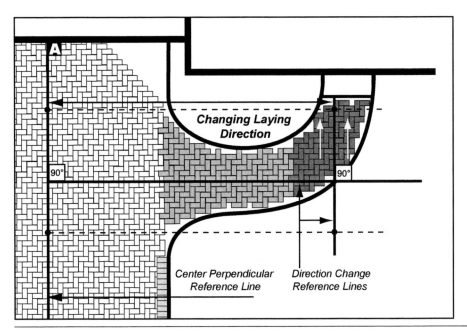

The same method is used again for the final direction change. This reference line is parallel to the center perpendicular reference line and 90˚ to our first direction change reference line. (Figure 19-7)

Changing Laying
Direction

90°   90°

Center Perpendicular
Reference Line

Direction Change
Reference Lines

Chapter 19 • Paving in Difficult Areas

20 Cutting

MARKING AND CUTTING PAVERS

Clean and accurate paver cuts are important to a professional paver job. The popularity of curves in hardscape design makes it necessary to cut pavers. These cuts need to be clean and accurate to maintain pavement design and bond lines. Cut pavers can also add to the design of the pavement.

Marking the cut

SOLDIER COURSE: *A border course where rectangular pavers are laid perpendicular to the edge restraint.*

Making use of tools to mark pavers in radius areas can make a slow and difficult job easy. The *QUICKDRAW* is a tool that makes mark a set distance from the edge. It is used for marking pavers **when** using a soldier course.

When laying pavers on a curved sidewalk with a soldier course, a field of pavers is laid first. The *QUICKDRAW* is used by setting one end on the edging. The marking line is drawn by placing one end of the unit on the edge restraint. The tool is set to mark the size of the soldier course paver plus a little extra (1/8 inch to 1/4 inch) to allow for final adjustment. An added feature is a wheel that is used on upright surfaces. (See Figure 20-3). Any type of marker can be used with the tool. It is important that the marking device be perpendicular to the edging or adjacent edges.

Marking pavers with a QUICKDRAW for a soldier course.
(Figure 20-1)

Marking pavers with a QUICKDRAW on an edge restraint.
(Figure 20-2)

Marking pavers around an object.
(Figure 20-3)

Bad sliver cuts make an irregular looking soldier course.
(Figure 20-4)

Even spaced alternating cuts make a smooth curve and less cutting per lineal foot.
(Figure 20-5)

Cutting pavers for a soldier course

A soldier course that follows a curve requires cutting. The one thing to keep in mind is to try to avoid sliver cuts. Figure 20-4 shows a very uneven soldier course with a number of unattractive sliver cuts.

Space full paver widths 1/2 inch from the outside radius.
(Figure 20-6)

When pavers are laid on a radius for a soldier course, not every paver needs to be cut. Start the radius with a full paver laid perpendicular to the edge. Lay the next full paver perpendicular to the edge, but a full paver width away from the previously laid paver on the outside edge of the radius. The space for the paver width should be measured 1/2 inch from the outside edge. Use full or cut pavers to get proper spacing.

Next, lay another full paver. Continue like this the full length of the radius.

Once the radius has been set with full pavers and left open spaces, place pavers on top of the full pavers. To support the paver temporarily while marking, place a small piece of paver under it.

Center the paver for marking over the opening. Cuts on both sides should be equal.
(Figure 20-7)

Set the pavers perpendicular to the edge. Carefully center them in the space.

Mark the cut lines on the paver. Cuts on both sides of the paver should be equal.

The finished curve will be smooth and even with fewer cuts.

Place a small piece of paver under the bottom of the paver for support while marking for cutting.
(Figure 20-8)

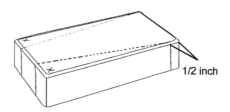

1/2 inch

Cuts should be equal on both sides without cutting into the last 1/2 inch at the widest end.
(Figure 20-9)

Cutting pavers with No soldier course

When a field of pavers directly abut a structure or surface, accurate cuts need to be marked on each paver.

The *PAVERSCRIBE* is a tool for marking pavers when **NOT** using a soldier course. This tool transfers the cutting angle and size to a paver for cutting. Place the tool in the space, using the adjustment lever to lock in the measurement. The tool is then set onto a paver for marking. This tool takes away the guess work when laying pavers around curved or straight surfaces without a soldier course and allows almost any skill level to mark just once for a perfect fit.

The PAVERSCRIBE in use on step work.
(Figure 20-10)

Set the PAVERSCRIBE to determine the angle and depth of the cut.
(Figure 20-11)

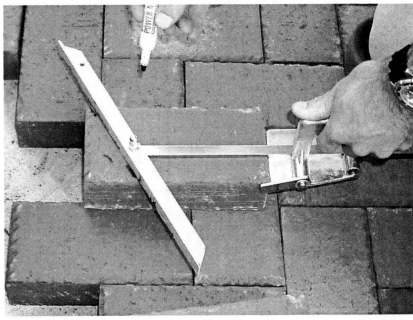

Transfer the angle and distance to a paver for cutting.
(Figure 20-12)

Chalkline on pavers.
(Figure 20-13)

Chalk Lines

When making straight cuts, a snapped chalk line works well. The difficulty with a chalk line is the possibility that the chalk will be rubbed or blown off. To temporarily maintain the chalkline on the paver, use clear lacquer and lightly spray over the chalk line. This will adhere the chalk to make the cuts, yet will be removed by cutting or during compaction of the joint sand.

Markers

Most individuals have a preference on the type of markers for the pavers. Some options are:

> *Permanent markers.* These work well to make a permanent mark. Cutting does not remove the mark; it remains visible.

> *Carpenter pencil.* These dull quickly when used on pavers. Also graphite is often difficult to see when cutting.

> *Soap stone.* Makes an easy to see white line that does not wash off or blow off but can be easily removed.

> *Construction crayons.* Generally good to use, will dull quickly, but marks stay while cutting.

Spraying a chalk line with lacquer prevents loss of the line prior to cutting.
(Figure 20-14)

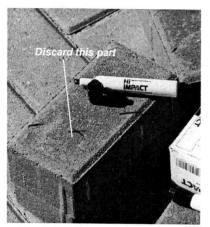

Discard this part

When marking pavers for cutting, it is a good idea to mark the part of the paver that will not be used. (Figure 20-15)

Different types of markers are permanent markers, marking crayons and soapstone. (Above) Carpenter pencil and sharpening a carpenter pencil with a utility knife. (Figure 20-16)

Markers (Figure 20-17)

Chapter 20 • Cutting

CUTTING OVERVIEW

The process of marking, cutting and fitting the pavers must keep up to the laying of the pavers. It is recommended that paver cutting stay not more than 5 feet from the laying edge on a residential job on a daily basis. The closer the better. This will lock the bond pattern and eliminate shifting from unexpected traffic.

Whether the paver is cut with a splitter or a saw, the paver is cut so that it drops into place easily and does not need to be hammered in. If the paver is forced in, it may damage surrounding pavers or force the bond lines to shift. Normally a 1/4 inch joint is allowed for cuts.

Sliver cuts can be acceptable to maintain paver patterns. The problem with sliver cuts is that during compaction the sliver can break or crumble.

When cutting pavers on a commercial project, architectural specifications may call for cuts no smaller than 1/3 paver. To accomplish this, often times the paver pattern needs to be changed at the edge. Make sure you explain this if no sliver cuts are requested.

There are two main types of paver cutters: splitters and saws. Clean and accurate cuts that maintain pavement design and bond lines are possible with either type.

Paver cuts on a curve (Figure 20-18)

Cut pavers defining a handicap parking area. (Figure 20-19)

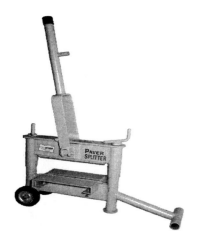

Handle position is too high for splitting pavers.
(Figure 20-20)

Paver splitters

Paver splitters are relatively maintenance free. Splitters use a concentric cam or hydraulically assisted to apply pressure to the blades to split a paver. They are less expensive and much quieter than masonry saws, but do not make as smooth of a cut as a masonry saw. Splitters produce almost no dust.

Splitters work fine for some jobs, but not all jobs. Paver splitters work well on tumbled or rough textured pavers. The paver splitter is a popular tool to rent for DIY projects. Splitters are also used on industrial jobs.

Always wear safety glasses when using a splitter.

How to use a paver splitter

- A paver splitter has both a top and bottom blade, adjustable height screws and a fixed angle table.
- Place the paver in the center of the table.
- Adjust the height of the top blade. The top blade should just touch the top of the paver when the handle is almost horizontal. This is the most effective position to exert pressure.
- The splitting blades are set into the paver with a few short strokes of slowly increasing pressure. This allows the blades to set into the surface of the paver. With a quick snap on the last stroke, the paver is split.
- The paver is split at a slight angle to create an undercut. This allows the top edge of the split paver to fit closer to the neighboring paver. (Figures 20-23, 20-24).

Correct handle position for splitting pavers.
(Figure 20-21)

Angled table makes for consistant and easy undercuts.
(Figure 20-23)

An undercut paver fits close.
(Figure 20-24)

Electric hydraulic splitter
(Figure 20-22)

Electric hydraulic splitter

A 12 volt electric hydraulic splitter is used in field for production use. It is used when a lot of splits need to be made quickly.

Saws

Masonry saws

Smooth and accurate cuts are made with masonry saws. When using any piece of power equipment, read and follow the manufacturer's operational instructions. Protect yourself with safety glasses, dust masks and approved hearing protection. Try to avoid cutting small pieces with the masonry saw. Cutting thin slivers of pavers any less the 3/8 inch (10 mm) can crack and be thrown from the saw. This can be dangerous to the individual cutting the pavers an well as the rest of the crew. It is usually possible to change the laying pattern or shift bond lines just enough to avoid having to cut very small pieces.

Masonry saws are available in gas or electric and table style or hand held models all using diamond blades. Place the saw on level, stable ground. When cutting a paver, hold it firmly against the stop plate of the saw's moveable table. Slowly slide the paver into and through the saw, cutting through the center of the cutting line. Pull the moveable table back to its starting position. Check to see that all markings are removed from the used side of the cut paver.

Nice sliver cuts.
(Figure 20-25)

Gas saws

The most common saw is a gas powered table saw. This type of saw is powerful and fast. One disadvantage is that they are heavy and they also require air filter maintenance. With constant use, this filter needs to be changed **daily**. To eliminate the cost of constantly changing the dry cartridge filters, consider an oil bath air filter conversion kit. This type of air filtration system keeps the motor running smooth and clean without the added expense of new filters every day.

Electric Saws

Electric table saws are also available. The advantage of an electric saw over a gas powered masonry saw is that they have much lower noise levels and are easy to turn on and off. The disadvantages of the electric table saw is if the saw draws 110-volt house power, it usually runs too slow.

Gas powered saw.
(Figure 20-26)

> Example: 1 to 1-1/2 hp electric = 110 volts
> 2 hp electric = 220 volts

If a 220 volt circuit hookup is not available on the project, a generator of at least 6,000 Watts will be needed to operate the saw.

Chop Saws

This type of saw is generally used for small jobs or repair situations only. They cut slowly and it is difficult to hold the paver while cutting. There are no water attachments, because the electric motor is not sealed. They are light and usually will not pop a circuit breaker in the house.

Electric masonry saw
(Figure 20-27)

Milwaukee chop saw
(Figure 20-28)

Hand held demolition saw.
(Figure 20-29)

Hand-held Gas Saw

The demolition saw is a staple of the construction industry. The many everyday uses of this hand held gas saw make it a must to have on every job site. It can be used for everything from cutting asphalt or concrete to rebar and tree roots.

When using the hand held gas-powered saw with pavers, some companies like its speed in cutting through in-place, marked pavers. Others like the convenience of cutting a few pavers without the tedious setup of a masonry saw.

However, dust is a problem with this kind of saw. There is no way to control the dust and if a water kit is used, the slurry will stain the pavers.

The disadvantages of this kind of saw is that they wear diamond blades faster and the potential for personal injury is greater.

Saw blades

Diamond Blades

A diamond blade is recommended for paver cutting. There are 2 types of diamond blades available. The less expensive blade has soldered or brazed segments of diamonds. The preferred blade is laser welded. Another difference in blades is the amount of diamonds in the cutting blade segments. Price is usually the indicator.

The sides of the blades wear faster than the edge. Watch that the segments do not wear too much on the sides. If the blade is worn on the sides, the paver will catch on the steel core and bind. The minimum recommended thickness for table saws is .110 inch. A thinner blade can be easily damaged or warped. The recommended blade thickness for a hand held gas saw is .125 inch.

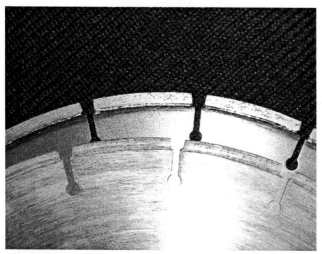

Worn masonry saw blade (front). New saw blade (rear).
(Figure 20-30)

New   Worn

Illustrated saw blade wear. Notice how the blade not only wears from the bottom but also on the sides.
(Figure 20-31)

Cutting Wet

The advantage of using wet masonry saws is that water traps the dust during cutting. Water also reduces wear and overheating of the diamond blade. When cutting wet, always feed fresh water to the saw for cutting, do not use recycled cut water (slurry). This will stain the pavers.

When cutting wet, the cutting area creates a very messy work area. The operator must be protected against caustic concrete spray and a water hook up is needed on the job site. It is important to control runoff and misting. This can easily damage and stain the new pavement, grass, shrubs, vehicles, and buildings.

Cutting Dry

Cutting dry requires less equipment. No water hookup is needed and there is less clean up. When cutting dry, a vacuum dust control system should be used. Before you begin, consider where the dust will settle to avoid staining your work or surroundings. Work downwind of the pavement if possible, and place tarps if necessary to block wind.

> *Note: Future safety regulations may ban dry cutting unless dust control systems are used.*

Blade Speed

The important rule when using a masonry saw is to maintain the blade speed when cutting. Do not feed the paver so fast that the blade slows. Feed the paver into the blade at a speed that does not slow down the blade. When the blade slows cutting efficiency greatly decreases.

Dust

Dust is a factor when cutting pavers. Controlling the dust can be done either by cutting wet or using a dust collection system. Some job sites require no dust. Dust can travel quite a distance and can coat vehicles, landscapes and buildings.

Dust control systems

Vacuum dust collection systems are more commonplace on jobsites as the only practical solution to control dust. They work by capturing the dust from the blade as the pavers are cut.

Dust control system
(Figure 20-32)

Dust collection hood.
(Figure 20-33)

Safety

When cutting pavers, always wear NIOSH (National Institute for Occupational Safety and Health) approved dust masks or respirators. The dust mask has a metal strip that fits to the nose. Always wear safety glasses when cutting or splitting pavers.

Concrete dust is harmful if workers are exposed to it without protection. Protection cream and gloves can protect workers when cutting. Always wash up with clean running water.

Safety glasses, air mask and ear protection
(Figure 20-34)

Protective skin cream
(Figure 20-35)

Ornamental paver cutting
(Figure 20-36)

Compacting and Setting Pavers

PAVER COMPACTION

Before paver compaction

Complete the pavement to be compacted. Adjust bond lines if necessary. Make and replace all cuts. Set an edge restraint on the open edge of pavement. Sweep all debris from the paved area.

Paver compaction starts the lock-up of the pavement. Final compaction is accomplished in two steps: the initial compaction forces the bedding sand up into the joints; with the final compaction, the sand is swept into the joints of the pavers and vibrated to compact.

Jointing sand

Joint sand is spread in the morning to dry on pavement. (Spread on the street or previously laid pavement.) Use a base rake to 'rib' the sand. This creates more surface area and will allow it to dry faster.

Using plate compactors

The type of installation and thickness of pavers used, determines how large a compactor to use. On a typical residential job, one capable of 3,500 to 5,000 pound (13 to 22-kN) of compaction force is acceptable. For pavers 3⅛ in. (80 mm) and thicker, use a minimum 4,000 pound plate compactor.

Textured surfaces may be damaged by a standard plate compactor. It is advised to use a special rubber pad or roller frame attachment on the compactor when the pavers have special textures.

As with any machinery, check it over before starting it up. Are the gas and oil levels good? Clean the air filter 1 to 2 times per day. Oil bath filters for equipment are more effective and less costly than typical cartridge filters.

The compactors will move ahead of the operator at its own rate of speed as it compacts. This is a normal function.

(Figure 21-1)

Compact the pavers — compact the sand

Start on the perimeter of the pavement. Follow around to the start, overlap the second pass by 4 in. to 6 in. Continue around, overlapping each round until the center. The initial compaction forces the bedding sand ¼ in. to 1 in. into the bottom of the joints. The operator walks behind the compactor watching for chipped or broken pavers; as he passes by a paver that will need replacing, the operator can mark it. Soapstone is a good paver marker.

Do not remove the pavers yet. The compactor must be far enough away. If the compactor is too close, sand will fill the void. When removing a paver, a Paver Extractor works best. Remove the paver and immediately drop the new one in its place.

Compact the pavement with lateral sweeps with the compactor back and forth across the pavement. A minimum of 2 full sets of compaction is necessary.

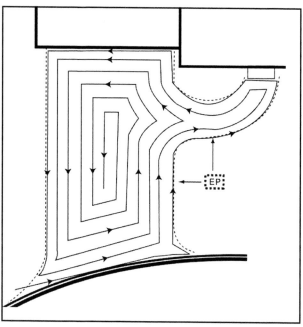

Step 1 – Perimeter compaction
(Figure 21-2)

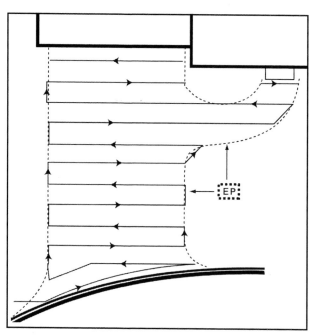

Step 2 – Lateral compaction
(Figure 21-3)

Sweep joint sand onto pavement

The second step to pavement lockup is filling the joints with jointing sand. Ideally, two people operate the brooms while a third operates the compactor.

Sweep dry joint sand onto the pavement with a stiff bristled push broom. The sand will fall into the joints as it is pushed around the pavement. Spread the sand evenly across the pavement.

The compactor is used in this step to vibrate the sand into the joints. Make lateral passes across the pavement. The vibration of the compactor will allow the sand to fill the joints completely. (Remember that while sweeping to leave enough sand behind to fill the champfers.) After compacting the entire pavement in one direction, crosshatch the pavement with a second set of passes. Keep the process going until no more open joints appear during compaction.

STAY AT LEAST 6 ft. AWAY FROM THE LAYING EDGE.

Water

If the bedding sand is saturated with water DO NOT COMPACT. Water drains quickly through sand. Wait until the bedding sand is dry prior to compacting the pavers.

Testing filled joints

After using the compactor to vibrate the joint sand until the joints are full of sand, give it the putty knife test. Try to insert a putty knife into the joints. Give the putty knife some hard pressure. If it does not slip into the joints, you have done well, and are finished. If it goes right in, pushing down sand with it... well, there is more work to be done. Sweep the joints again with sand and vibrate the pavement. Knife it again and if it still fails the test, keep filling, vibrating, and testing until it passes.

Forgiveness

The final compaction and filling the paver joints is the only opportunity to achieve a perfectly smooth pavement. Once the joints are filled, the interlock has begun.

End of day

- Sweep any excess dry joint sand into a pile and cover it with a tarp.
- Cover the uncompacted, unrestrained edges with canvas or plastic if the weather is going to turn bad.

Joint sand stabilization

Organic or chemical stabilization, wet or dry

It is sometimes necessary or desirable to stabilize joint sand to prevent loss due to water and wind erosion. Sandlock is an organic product that binds the joint sand when water is applied.

When using Sandlock with joint sand, it is recommended that mixing be done in a mixer on site. Because Sandlock binds particles together, it is not recommended to mix on asphalt, as the oils in the asphalt are drawn into the mixture.

On site mixing ratios can be adjusted between 1½ to 2 pounds of Sandlock for every 100 pounds of joint sand. Mix the Sandlock and sand in a mixer. Sweep the joint sand into the joints as instructed and compact.

It is important to cleanly sweep all the mixture off the pavers when compaction is complete. Then, use an electric or gas blower to clean off any remaining Sandlock. Activate the material with water, using a light spray from a hose to flood the surface. Do not pressure wash; a pressure washer could remove joint sand.

A Sandlock stabilized pavement will also prevent excessive plant growth and ants. Any loosened material will rebind with any moisture, but will stay intact in excessive rain or wind.

Chemical stabilization

Joint sand can also be stabilized with chemical treatments. This is usually a second process after the pavement is completed. This process usually leaves some material on the surface, which may or may not be desirable.

Glossary

The following are terms used in the production, design, construction and testing of interlocking concrete and brick pavements.

ABSORPTION – Weight of water incorporated by a concrete paver unit during immersion under prescribed conditions, expressed as a percentage of the dry weight of the unit.

ABRASION – The mechanical wearing, grinding, scraping or rubbing away (or down) of paver surfaces by friction or impact, or both.

ADHESION – A soil property which causes the soil to stick to the contact surface of the machine

ADMIXTURE – Prepared chemicals added to the concrete mix immediately before or during the mixing of water, cement and aggregate to improve density, durability and strength.

AGGREGATE – Crushed gravel, sand or rock used in road surfaces, concrete or asphalt mixes.

AMPLITUDE, ACTUAL – The vertical distance a vibrating drum or plate moves from its rest position during one vibration

ASPECT RATIO / PLAN RATIO – The overall length of a paver divided by its thickness. A ratio of 4:1 is the maximum recommended for vehicular pavement applications. Also see Plan Ratio.
Example: A 4 in. (100 mm) wide, by 8 in. (200 mm) long, by 3⅛ in. (80 mm) thick paver has an aspect ratio of 2.5.

ASTM – American Society for Testing and Materials

BASE COURSE – A material of a designed thickness placed on a sub-base or a subgrade to support a surface course. A base course can be compacted aggregate, cement or asphalt-stabilized aggregate, asphalt, concrete, or flowable fill.

BEDDING SAND – A layer of coarse, clean, sharp sand that is screeded smooth for bedding the pavers. The sand can be natural or manufactured. Example: crushed from larger rocks, and should conform to the grading requirements of ASTM C-33. This layer is 1 in. (25 mm) thick.

BEDDING SAND is a vital component of the flexible paving system.

BEDDING SAND DEGRADATION TESTS – Evaluation of the degree of attrition of fine aggregate usually with steel balls or other abrading devices agitated with sand in a container. Pre-and post-testing sieve analysis are conducted to determine the increase in fines. The tests are used to evaluate the durability of bedding sand under heavy loads or channeled traffic.

FULL RANGE PAVERS in cubes.

CHAMFERS show in a variety of styles on clay and concrete pavers.

BENTONITE CLAY – A clay with a high content of the mineral montmorillonite, usually characterized by high swelling on wetting that can be used to help seal paver joints.

BINDER – Fines which fills voids and hold gravel or other aggregates together.

BLENDED OR FULL RANGE PAVERS – Mixing of colored concrete pavers from three or four cubes to insure even color distribution. Pavers can be manufactured in a variety of colors.

BLIND SPACER – A spacer that stops short of the paver top surface. This kind of spacer cannot be seen after the pavers have been installed and the joints are filled with sand.

CAPILLARY – The soil channels which allows water to be moved vertically or laterally through soil.

CBR – The California Bearing Ratio provides an index of strength for the subgrade soil which has been measured for its ability to take loads under a pavement structure. It is expressed as a percentage of the load required to penetrate crushed aggregate road base material.

CEMENT – AGGREGATE RATIO – The proportional weight of cement to fine and coarse aggregate in concrete.

CEMENT, PORTLAND – A hydraulic cement produced by pulverizing clinker consisting essentially of hydraulic calcium silicates, and usually containing one or more forms of calcium sulfate.

CENTRIFUGAL FORCE – The force generated by the vibration-inducing mechanisms at a stated frequency.

CHAMFER – 45 degree beveled edge around the top of a paver unit, usually 1/16 to 1/8 in. (2-3 mm) wide. It helps water drain from the surface, facilitates snow removal, helps to prevent chipping and spalling, and delineates the individual pavers.

CLAY – A cohesive soil made up of decomposed rock and microscopic fines, with putty-like properties. It is plastic (sticky) to the touch when wet and exhibits considerable strength when air-dry.

COHESION – The property of soil that holds its particles together.

COLOR BLEND – A paver with two or more colors created by blending pigments during the manufacturing process to produce a variegated appearance.

COMPRESSIBILITY – Property which causes soil to deform under stress of a weight bearing load.

COMPRESSIVE STRENGTH – The measured maximum resistance of a concrete paver to loading, expressed as force per unit cross-sectional area (in pounds per square inch).

PAVE TECH ©

CONCRETE GRID PAVERS – Concrete units that have up to 50 percent open area. They are usually a maximum of 16 by 24 inches. The openings can be filled with turf or aggregate to promote infiltration of storm water. They are usually used for intermittent parking areas, access lanes, or for erosion control on embankments.

CONCRETE SAND – A coarse, washed sand conforming to the gradation requirements of ASTM C-33.

CREEP – Slow, lateral paver movement from horizontal forces such as braking tires. The movement is usually imperceptible except to observations over a long duration.

CROWN – Elevated center of a pavement. It is beneficial to paver surface draining and interlock.

CRUSHED STONE – A product used for pavement bases that is made from mechanical crushing of rocks, boulders, or large cobblestones at a quarry. All faces of each aggregate have edges that result from the crushing operation.

CRUSHER-RUN – The total unscreened product of a stone crusher.

CSA-A231.2 – Canadian Standards Association (CSA) product standard for precast concrete pavers (interlocking units) that gives standards for dimensions, minimum compressive strength, and durability under freeze-thaw cycles with deicing salt through various test methods.

CUBE – Factory stacked pavers which are strapped or wrapped, with or without a pallet for shipping to a site. The number of pavers in a cube varies with their thickness and shape.

Banded CUBE

DEFLECTION – A temporary movement of the pavement structure due to traffic loads.

DEFORMATION – A change in the shape of the pavement.

DENSE GRADED AGGREGATE – Uniform grade from maximum size to minimum size particles. Compacted specimen has very few voids.

DENSITY – The weight of a unit of material as compared to its volume – pounds per cubic foot (kg/m). Usually expressed in pounds per cubic foot. Density describes how closely the particles of a soil are compacted. Also, density in reference to paver units themselves is the mass per unit volume.

DENTATED PAVERS

DENTATED PAVER – A paver that is not rectangular or square in shape.

DRAINAGE COEFFICIENT – Factors used to modify layer coefficients of pavements. It expresses how well the pavement structure can handle the adverse effect of water filtration.

EDGE PAVER – A paving unit that is made with a straight, flush side, or cut straight for placement against an edge restraint.

EDGE RESTRAINT – A curb, edging, building or other stationary object that contains the sand and pavers so they do not spread and lose interlock. They can be exposed or hidden from view.

Pave Edge® EDGE RESTRAINT

EFFLORESCENCE on pavers

FALSE JOINTS - concrete

FALSE JOINTS - clay pavers

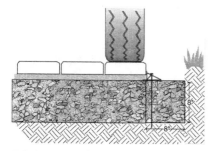

FLEXIBLE PAVEMENT SYSTEM

EFFLORESCENCE – A naturally-occurring white deposit of calcium carbonate that can form on any concrete surface. It results from the reaction of calcium hydroxide with carbon dioxide from the air and is a by-product of cement hydration. It does not affect structural integrity and will dissipate over time. Efflorescence is not indicative of a flawed product. If desired, it may be removed by washing the surface with a special paver cleaner.

ELASTICITY – Soil property which allows soil to return to its original configuration when the compression force is removed.

EQUIVALENT SINGLE AXLE LOADS (ESALS) – Summation of equivalent 18,000 lb. force single-axle loads used to combine mixed traffic to design traffic for the design period.

EQUIVALENT THICKNESS – A comparison of the concrete paver and sand bed layer to the required thickness of asphalt to achieve an equivalent strength. A conservative design assumes this to be 6½ in. (165 mm)

EXPANSION – Increase in material volume due to increase in moisture content.

FALSE JOINTS – False joints or dummy grooves in pavers give the appearance of multiple individual paver units. They contribute to the installed pattern and can enhance the beauty of the pattern

FINES – Small clay or silt particles.

FLAG – See Paving Slab.

FLEXIBLE PAVEMENT – A pavement structure which maintains contact with and distributes loads to the subgrade. The base course materials rely on aggregate interlock, particle friction, and cohesion for stability.

FLEXIBLE SEGMENTAL PAVING SYSTEM – The flexible segmental paving system has four main components:
1. A uniform thickness compacted **GRANULAR BASE**. [The GRANULAR BASE typically consisted of crushed graded aggregates from 3/4 in. to fine dust and all sizes in between. The base is compacted in uniform layers to a thickness determined by the pavement use and existing soils. A minimum of 4 in. for patios and walkways and 6 in. to 8 in. for driveways. The purpose of the compacted granular base is to provide the major component of pavement strength and load handling.]
2. **SAND BEDDING** layer. [The BEDDING SAND layer is installed between the compacted aggregate base and the pavers. It must be uniform in thickness between 1 in. to 1½ in. of loose screeded sand. 1in. is preferred. The most common sand used for the sand bedding layer is coarse washed concrete sand. Check with your supplier because different areas of the country have varied definitions for coarse washed concrete sand. The purpose of the sand bedding layer is the load transfer layer and also to allow moisture movement from above and below. When installing the sand bedding layer do not use stone dust. Stone dust contains too many fines and when compacted does not allow for moisture movement.]
3. **EDGE RESTRAINT** [EDGE RESTRAINTS are critical for the pavement to perform. The edge restraints are necessary to maintain interlock and prevent paver or bedding sand shifting causing failure.]
4. **PAVERS** that act as a wear course and help spread loads over a larger area. [PAVERS act as the main wear coarse of the flexible paving system. The shape, thickness, bond pattern and size are all important components that add support to the interlocking pavement system.]

FLEXURAL STRENGTH – The property of a concrete paver that indicates its ability to resist failure in bending.

FLOWABLE FILL – A self-leveling, low-density cement like backfill material that attains 100% compaction without tamping or vibrating. It replaces compacted soil or conventional backfill as a structural fill that drains and is no stronger than the surrounding soil after it has obtained its ultimate strength.

FREQUENCY – The number of complete cycles of the vibrating mechanism per minute. Also called Hertz (HE).

FROST ACTION – Freezing and thawing of moisture in pavement materials and the resultant effects on them.

FROST HEAVE – The raising of a pavement surface due to water infiltration, which when allowed to accumulate, forms into ice in freezing climates.

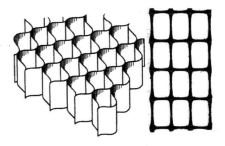

GEOGRIDS – three dimensional (left) and two-dimensional.

GEOGRIDS – Geogrids are two dimensional or three dimensional. The two dimensional type are flat and have small, television-screen shaped openings. The material is generally placed between the soil and the base to reduce rutting. Three-dimensional geogrids are 8 in. high and provide stability under loads for cohesionless soils.

GEOTEXTILE – A polypropylene material that is available in woven, nonwoven and grid styles. Used to keep separate or contain parts of the segmental pavement and to reduce rutting.

GRADATION – Soil or aggregate distributed by mass in specified particle-size ranges. It is usually expressed in percent of mass of sample passing a range of sieve sizes. Different sizes of aggregate.

GRADE – Surface elevation or slope.

GRANULAR BASES – Crushed or quarried stone material that generally ranges in size from 3/4 in. (19 mm) maximum for crushed stone to 1½ in. (38 mm) maximum for quarried stone down to small sand-sized particles. The sizes must be uniformly graded in various proportions that create a dense material when compact.

HARD FACING on a paver

GRANULAR SOIL – A particular type of soil, usually sand or gravel where the particles do not stick together.

HALF STONE – One part of a paver split in two equal parts.

HARD FACING – Dense top coat of hard fine aggregates.

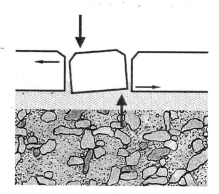

ROTATIONAL INTERLOCK *keeps pavers from shifting*

VERTICAL INTERLOCK

IMPERVIOUS – Resistant to the penetration of water.

INTERLOCK –
Interlock in segmental pavements is a result of the interaction of the pavers, edge restraint, and jointing sand. Interlock is a force that transfers loads in a segmental pavement. Interlock forces are horizontal, rotational and vertical load transfers. A downward load causes movement, and movement causes transfer.
HORIZONTAL INTERLOCK in the main body of the pavement prevents pavers from horizontal movement. This is achieved with bond pattern, adequate compaction of sand between the pavers and the friction between the sand and pavers.
ROTATIONAL INTERLOCK assures that there is no rotational movement of the pavers. If a load is applied asymmetrically to a paver, interlock holds the paver in place when there is a proper edge restraint and compaction. Rotational interlock is aided by having sufficient paver thickness. Recommendations for pavers in different applications to maintain rotational interlock are:

     Walks, patios and driveways 2 in. (50 mm) +.
     Streets and commercial sites 2⅝ in. (70 mm) +.
     Heavy industrial 3⅛ in. (80 mm) +.

VERTICAL INTERLOCK is achieved with proper sand between the pavers. By filling and compacting the paver joints sufficiently vertical interlock keeps pavers from movement. Friction with the sand will transfer the load to surrounding units.
PERIMETER INTERLOCK is the interface of the segmental pavement and surrounding landscape or hard surface. Perimeter interlock permanently restrains the edge of a segmental pavement. The edge restraint can be an existing hard surface, a purpose designed edge restraint or a poured or precast curb.

JOINT SAND – Sand swept into the openings between the pavers.

JOINT SPACING – The distance between pavers subsequently filled with joint sand.

LAYING FACE – The exposed, vertical row of pavers on bedding sand.

LAYING PATTERN – The repetitive geometry created by the installed units. Laying patterns may be selected for their visual or structural benefits.

LIFT – A layer of fill as spread or as compacted. Also a measurement of the depth of the material. The rated soil depth of compaction a compactor can achieve.

LIMESTONE SCREENINGS – A residual by-product of crushed rock containing particles small enough to pass a No. 200 sieve. It is not suitable for use as bedding sand.

LOCK UP – The initial settling period of concrete paver installations which progressively stiffen under traffic to a point that no further settling will occur unless failure develops in the base or subgrade.

MACRO TEXTURE – The deviations of a pavement surface from a true planar surface with dimensions generally from 0.5 mm and up or those that no longer affect tire-pavement interaction.

MASON SAND – A fine washed sand conforming to the gradation requirements of ASTM C-144.

MECHANICAL OR MECHANIZED INSTALLATION – The use of specially designed machines to lift and place layers of pavers onto screeded bedding sand in the final laying pattern. It is used to increase the rate of paver installation.

MICRO TEXTURE – The deviations of a pavement surface from a true planar surface with dimensions generally less than 0.5 mm.

MODIFIED PROCTOR TEST – This test is a variation of the Standard Proctor Test used in compaction testing which measures the density-moisture relationship under a higher compaction effort.

MODULUS OF ELASTICITY – The ratio of stress to strain for a material under given loading conditions.

MOISTURE CONTENT – The percentage by weight of water contained in the pore space of soil, sand or base, with respect to the weight of the solid material.

MORTAR – Cement and lime, or masonry cement, mixed with fine sand.

MULTI-COLORED PAVER (COLOR BLEND) – A paver with two or more colors.

NUCLEAR DENSITY TESTING – A method to accurately test soil density or moisture with a device that utilizes a probe inserted into the soil that emits radioactive rays which are measured by a Geiger counter.

OPEN-GRADED AGGREGATE – An aggregate that has a particle-size distribution, when it is compacted, has relatively large spaces between them. It can be used as a drainage course in base design.

OPTIMUM MOISTURE CONTENT – The water content at which a soil can be compacted to a maximum dry unit weight by a given compactive effort.

PADFOOT (SHEEPSFOOT) – Specially designed and arranged projections on a compactor drum that create a manipulative force.

PAVER EXTRACTOR – A tool used to grab a paver and remove it from the laying pattern.

PAVER EXTRACTOR

PAVER SPLITTER (Guillotine Splitter) – A hand operated machine, sometimes hydraulically assisted, for cutting pavers.

PAVING SLAB (or Flag) – A paving unit larger than an interlocking concrete paver. Maximum dimensions are generally 36 by 36 inches. Unlike concrete pavers, paving slabs to do not rely on interlock as the primary means of load distribution.

PAVER SPLITTER

CLAY PAVERS come in various sizes for different applications

CONCRETE PAVERS 2 3/8" (6 cm), 3 1/8" (8 cm), and 4" (10 cm)

PAVER THICKNESS AND HEIGHT – Pavers used in flexible systems typically vary in thickness from approximately 2 in. to 4 in. (50-100 mm).

Less than 2 inches (50 mm)
• Recommended for pedestrian traffic only.
• 1¼ in. pavers are too thin for sand set use. They are normally only used in mortared applications.

More than 2 inches (+50 mm)
• Recommended for walks, patios, driveways, and plazas.
• Recommended for commercial, industrial, roads, and parking lots.

More than 4 inches (100 mm).
• Recommended for extremely heavy traffic.
• Heavy industrial use such as container ports or mining.

PAVEMENT PERFORMANCE – The trend of serviceability under repetitive loads.

PAVEMENT REHABILITATION – Work done to an existing pavement to extend its service life. This includes placement of additional surfacing material and/or other work necessary to return an existing roadway to a condition of structural or functional adequacy. This could include the complete removal and replacement of the pavement structure.

PAVEMENT STRUCTURE – A combination of subbase, base course, and surface course placed on a subgrade to support the traffic load and distribute it to the roadbed.

PERFORMANCE PERIOD – The period of time that an initially constructed or rehabilitated pavement structure will last before reaching its terminal serviceability. This is also referred to as the design period, expressed in years. Twenty years is normally used in North America.

PERMEABILITY – Property of material which permits water to flow through it.

PERMEABLE INTERLOCKING PAVEMENT – Concrete pavers with wide joints (10 mm to 30 mm) or a pattern that creates openings where rain can infiltrate. The openings are filled with aggregate or topsoil and grass. The pavers are typically placed on an open-graded aggregate base that stores runoff.

PLASTIC – The ability of soil with some water content to be rolled into thin threads.

PLATE COMPACTOR – Also known as a plate vibrator; which is used to compact pavers into bedding sand in order to promote interlock among the individual units.

PLASTIC LIMIT – 1) The water content corresponding to an arbitrary limit between the plastic and the semisolid soil consistency. 2) Water content, measured by soil just beginning to crumble when rolled into a thread approximately 1/8 in. (3.2 mm) in diameter.

POZZOLANIC MATERIALS – Fly ash, pozzolan, silica lime, or blast furnace slag used as cement substitutes. They are used in the concrete mix to increase concrete paver density and durability.

PREPARED ROADBED – In-place roadbed soils compacted or stabilized according to provisions of applicable specifications.

PRESENT SERVICEABILITY INDEX (PSI) – A rating, usually between zero (completely nonfunctional) and five (new/perfect), that measures pavement conditions. This convenient, general rating method measures a pavement's overall condition and usefulness over time.

PROCTOR COMPACTION TEST – A test created by R.R. Proctor which measures the relationship of soil density with respect to soil moisture content under a standard compaction effort. This test identifies the maximum density obtainable at an optimum moisture content of a particular material.

PROGRESSIVE STIFFENING (LOCK-UP) – The tendency of pavements to stiffen over time. Interlocking concrete pavement stiffens as it is receives increasing traffic loads.

PUMPING – The ejection of bedding and joint sand, either wet or dry; through joints or cracks, or along edges of pavers when a load is applied.

RIGID BASE – Reinforced or non-reinforced concrete slab on grade.

RUTTING – Permanent deformation from repetitive traffic loading that exceeds the ability of the pavement structure to maintain its original profile.

SAND – Non-cohesive granular material usually under 3/16 in. (5 mm) in size, made from the natural erosion of rocks, and consists of sub-angular or rounded particles. Sands made by crushing of coarse aggregates are called Manufactured Sands.

SAILOR COURSE – A single or double border course of rectangular pavers laid parallel (lengthwise) to the edge restraint

SCREED BOARD (Strike Board) – A rigid, straight piece of wood or metal used to level bedding sand to proper grade by pulling it across guides or rails set on the base course or edge restraints.

SCREED GUIDES – Grade guides, such as pipe, that will guide the screed in producing the desired elevation of the bedding sand.

SCREENINGS – A residual product not suitable for bedding sand. It is a by-product from the crushing of rock, boulders, cobble, gravel, blast-furnace slag or concrete. Most of the aggregate passes the No. 4 (4.75 mm) sieve.

SEALER – A material usually applied as a liquid that is used to waterproof, enhance color, or stabilize joint sand in interlocking concrete pavements.

SEGMENTAL PAVEMENT – Segmental paving is the implementation of flexible paving with clay or concrete pavers. The paving system must contain a compacted aggregate base, a sand bedding layer, wear course (pavers) and edge restraint.

SEMI-RIGID BASE – Asphaltic concrete commonly referred to as asphalt.

SHRINKAGE – Reduction in volume when moisture content is reduced.

SIEVE ANALYSIS – Determines partical size within a particular sample of soil.

SILT – Material passing the No. 200 (75 µm) U.S. Standard sieve. A heavy soil intermediate between clay and sand.

SKID RESISTANCE – A measure of the frictional characteristics of a surface with respect to tires.

SLUMP – A measure of consistency and water content in freshly mixed concrete. Slump is measured from a specimen immediately after removal of a cone shaped mold. Unlike ready-mixed concrete, pavers are zero slump concrete because of low water content. They are not tested for slump.

SOIL SEPARATION FABRIC – A layer of fabric typically placed between the subgrade and the base to reduce rutting, also called a geotextile. See geotextile.

SOIL STABILIZATION – Chemical or mechanical treatment designed to increase or maintain the stability of a mass of soil or otherwise to improve its engineering properties. Lime, fly ash or cement are typical chemical stabilization materials. Geotextiles and geogrids are typical mechanical materials for soil stabilization.

SOLDIER COURSE – A border course where rectangular pavers are laid perpendicular to the edge restraint.

SOLID COLOR – Pavers having only one color.

SPACERS, SPACER BARS, RIBS OR NIBS – Small protrusions molded into the sides of pavers during manufacturing to keep them uniformly spaced during installation. Spacers help prevent chipping and spalling.

SPALL – A flake-like fragment that detaches from the edge or surface of a brick or paver by a blow, from severe weather or pressure from adjacent units. This is minimized with concrete pavers and some clay pavers due to chamfers and spacer bars.

STABILIZED BASE – An aggregate base where either cement, asphalt or other material is added to increase its structural capacity The soil subgrade can be stabilized with cement, lime, fly ash or other materials.

STACK BOND – A laying pattern in which the joints in both directions are continuous.

STATIC FORCE – Force exerted on soil or asphalt by the machine's static weight.

STATIC WEIGHT – The operating weight of the machine exerted on the material while the machine is at rest.

SUBBASE – The layer or layers of specified material of designed thickness placed on a subgrade to support a base course.

SUBGRADE – The existing site soil upon which the pavement structure is constructed.

TENSILE STRENGTH – Maximum unit stress which a paver is capable of resisting under axial tensile loading, based on the cross-sectional area of the paver before loading.

TEXTURED OR ARCHITECTURAL FINISH – Aesthetic finishes for paver surfaces such as sand blasting, bush hammering, tumbling, grinding, polishing, or washing.

TOP COAT (or Hard Facing) – Applying a thin layer of fine aggregate and cement to the top surface of a concrete paver The layer is often colored and is used to provide a more intense appearance, greater abrasion resistance, or provide a base for a textured finish.

TOPSOIL – Surface soil, usually containing organic material.

WATER-CEMENT RATIO – The weight of water, divided by the weight of cement in a concrete mixture. Concrete pavers usually have a water-cement ratio of 0.27 to 0.33, lower than ready-mix concrete, which contributes to strength and durability.

ZONING – Using different paver colors, textures, shapes, patterns, and surface elevations to delineate pedestrian and vehicular areas.

ENGINEERING TERMS USED IN DESIGN OF CONCRETE INTERLOCKING PAVERS

Terms Used In Architectural Design And Detailing Of Interlocking Concrete Pavement

BISHOP'S HAT – A five-sided paver often used as an edge paver with a 45 degree herringbone pattern.

BASKET WEAVE (PARQUET) – A paver pattern where two or move pavers are placed side-by-side. Adjacent pavers are placed side-by-side, but turned 90 degrees and alternated 90 degrees throughout the pattern.

ENGRAVED PAVERS – Pavers engraved with letters or images by molding during manufacture, shot blasting, or have a cast metal plate inset into the surface.

HARD EDGES – Pavers set against a visible edge restraint to visually reinforce the pavement edge.

HERRINGBONE PATTERN – A pattern where joints are no longer than the length of 1½ paver. These patterns can be 45 degrees or 90 degrees depending on the orientation of the joints with respect to the direction of the traffic.

HUMAN SCALE – Using paver sizes, patterns, colors and textures next to large buildings or open areas. It reduces the perception of the large scale of the spaces.

MARKERS – Marking underground utilities, traffic direction, parking stalls, lanes, pedestrian/vehicular areas, or other areas with pavers of different colors, textures or shapes.

MOSAICS – Paving uses as pictorial maps, murals, or geometric patterns as a landmark, to emphasize an area or to suggest movement.

PARQUET – See Basket Weave.

REFLECTING PATTERNS – Using Pavers to mirror geometric patterns, shapes, colors or textures in the surrounding site.

RUNNING BOND COURSE (SAILOR COURSE) – A paver course or two where lengths abut against the edge restraint.

RUNNING BOND (STRETCHER BOND) – A paver pattern with continuous joint lines in one direction. Pavers are staggered from one row to the next.

SAILOR COURSE – See Running Bond Course.

SLIP RESISTANCE – Resistance against pedestrian slipping, defined as the ratio of a minimum tangential force necessary to initiate sliding of a pedestrian's shoe or related device over a surface. Non-mobility impaired persons require minimum coefficient of friction values ranging from 0.2-0.3. Wheelchair users require friction values ranging fro 0.5-0.7. Crutch users and those with artificial limbs require values from 0.7 to 10. Clean concrete pavers generally have values exceeding 0.7.

SOFT EDGES – Paver with no visible edge restraint that, usually meets grass or other vegetation, giving a soft appearance at the edge.

SOLDIER COURSE – A Paver course where widths abut against the edge restraint.

STACK BOND – A laying pattern in where the joints in both directions are continuous.

STRETCHER BOND – See Running Bond.

TACTILE PAVERS – A paver detectable by sight-impaired persons due to change in color or texture from surrounding surfaces. Changes in texture are achieved with detectable warnings.

References

Canadian Standards Association, <u>Precast Concrete Pavers</u>, CSA-A23 1.2-95, Rexdale, Ontario, 1995.

Aggregate Producers Association of Ontario, <u>Construction Aggregate Consumers' Guide</u>, Downsview, Ontario, 1990.

American Association of State Highway and Transportation Officials, <u>Guide of the Design of Pavement Structures</u>, 1993, Washington, D.C.

American Society of Testing and Materials, <u>Annual Book Of ASTM Standards</u>, Vols. 4.02, 4.03, 4.05, 4.08, 1991, Philadelphia, PA.

Concrete Paver Institute, <u>Building Interlocking Concrete Pavements, A Basic Guide</u>, 1991, Herndon, VA.

Pavers by Ideal, Larry Nicolai, <u>A Contractor's Guide to Installing Interlocking Concrete Pavers</u>, 2002, Westford, MA.